AF797644

TEACHING HISTORY WITH CHATTY GEEPS

TEACHING HISTORY WITH CHATTY GEEPS

A TECHNOCURIOUS APPROACH TO USING GENERATIVE AI IN THE CLASSROOM

AMY ALLEN
DAVID HICKS

VIRGINIA TECH PRESS

VIRGINIA TECH PRESS
BLACKSBURG, VA

Suggested citation: Allen, A., & Hicks, D. (2026). *Teaching history with Chatty Geeps: A technocurious approach to using generative AI in the classroom*. Virginia Tech Press. https://doi.org/10.21061/chattygeeps

Publisher: This work is published by Virginia Tech Press.

Virginia Tech Publishing & Press
University Libraries at Virginia Tech
560 Drillfield Drive
Blacksburg, VA 24061
https://publishing.vt.edu | publishing@vt.edu

Peer review: This book has undergone single-anonymous peer review by three reviewers.

Accessibility statement: Virginia Tech Publishing & Press is committed to ensuring that its publications are accessible to a wide range of users and comply with the Americans with Disabilities Act, including the 2024 Title II regulation requiring web content to meet WCAG 2.1 Level AA standards. Our publications are produced in multiple formats—including HTML, PDF, print, and EPUB—with accessibility features such as structured tagging, alternative text, and navigable content to support a wide range of assistive technologies. If an image requires long description in addition to alternative text, the long description can be found at the end of the chapter or by clicking directly on the image. Virginia Tech Publishing & Press is continuously working to improve accessibility and welcomes any feedback from readers.

Use of AI: This publication was developed with the use of generative artificial intelligence.

Cataloging-in-Publication Data
Allen, Amy, & Hicks, David
Teaching History with Chatty Geeps: A Technocurious Approach to Using Generative AI in the Classroom / Amy Allen and David Hicks
Pages cm
ISBN 978-1-962841-56-6 (PDF)
ISBN 978-1-962841-55-9 (EPUB)
ISBN 978-1-962841-57-3 (Pressbooks; https://pressbooks.lib.vt.edu/chattygeeps)
ISBN 978-1-962841-54-2 (Print)
URI: https://hdl.handle.net/10919/142386
DOI: https://doi.org/10.21061/chattygeeps

Artificial intelligence—Educational applications.
History—Study and Teaching.
Generative artificial intelligence.
LB1028.44 .A45 2026

Cover images: Public domain
Cover art: Hillary Meadows
Cover design: Madison Spears

CONTENTS

FOREWORD

Jeremy D. Stoddard, University of Wisconsin

> Books will soon be obsolete in the schools...scholars will soon be instructed through the eye. It is possible to touch every branch of human knowledge with the motion picture.
>
> —Thomas Edison, 1913 (Cuban, 1986, p. 11)

> The one requirement for a good and a universal education is an inexpensive and readily available means of teaching children. Unhappily, the world has only a fraction of the teachers it needs. Samoa has met this problem through educational television.
>
> —Lyndon B. Johnson, 1968 (Cuban, 1986, p. 30)

For over a century, every new technology with a potential for educational use has been viewed as a replacement for a highly trained educator. Artificial intelligence (AI) and more recently generative AI (GenAI) are no exception, having led to similar proclamations that teachers will soon be replaced by smart tutors and individualized education powered by large language models (LLMs) trained to support content acquisition and skill development aligned with state academic standards and assessments. What these references to educational technologies all have in common is that they envision learning as an acquisition of facts and narratives—knowledge—and not as the social production of knowledge through critical and systematic analysis. They view learners as people who need to learn discrete and testable facts and skill sets and not as members of a community whose goal is to develop skills and experience in collaborative problem solving, deliberation, and consensus building, in order to plan for how to build the world they want to live in.

Full disclosure, I am not an AI early adopter. In fact, I have long been skeptical of AI applications in education and in media platforms on critical and ethical grounds. Critical as there are clear limitations to the use of current AI platforms that greatly hinder their application in effective teaching and learning. Perhaps more importantly, there are ethical dilemmas associated with the use of GenAI in particular, from the environmental impact of their energy and water use to copyright issues with the material they are trained on.

Pedagogically, there are many technical and ethical concerns in how GenAI models are being applied in education. These concerns include the hallucinations in AI responses and summaries as well as applications such as Character.AI that invite users to chat with bots impersonating historical figures (see Chapter 11). Take, for example, my interactions with a Character.AI bot impersonating General George S. Patton, who became famous during World War II for his bold victories and for his questionable demeanor, actions, and "salty" language.

> Me: *If you knew how actions such as slapping a shell-shocked (aka PTSD) soldier in North Africa or calling a Soviet counterpart general a "son of a bitch" would impact your career, would you still have done it?*
>
> Patton: *I wish I'd "kissed" the son-of-a bitch.*

GenAI models such as ChatGPT, Gemini, and Claude are trained on a vast amount of data available on the open web and even on copyrighted materials accessed illegally (including one of my books). The quote above is from the film *Patton* (1970), and lists of quotes from the film are widely available online and become part of this data set. If I asked my chatbot Patton what he knew about my work he might respond "You magnificent bastard, I read your book!" (another movie quote).

That is to say the response you get from a chatbot Patton or even a prompt to a GenAI model is going to be based on the source material it was trained on; the natural language models and machine learning it was programmed on; and any intentional programming its designers chose, including programming to make the interactions as convivial and human-like as possible. This means that questions about history will be based upon a wide range of sources with varying degrees of historical accuracy and evidence, from historical fiction to books written by historians.

There are also ethical issues with how LLMs are trained and with the data used to train them. In addition to illegal scraping of mass quantities of data with copyright protections, the web is also flush with data that reinforces inequalities in media representation and in society in general (Benjamin, 2019). These issues with GenAI are distinct from other roles that AI have played in society for a long period of time. More tightly trained AI models have been used, for example, to more effectively screen for different forms of cancer. These more focused goals of purpose-built AI models differ greatly from the personal assistant and do anything goals of GenAI.

That brings us to this incredibly engaging and useful book by Amy Allen and David Hicks. This book thoughtfully and playfully wrestles with these technical, pedagogical, and ethical issues surrounding the role that GenAI can play in history education. This book helps us to consider both the affordances and limits of GenAI for teaching and learning history as well as strategies for teaching history *with* and *about* AI. In particular, the authors focus on the many complexities of using AI models in relation to history teaching and learning through inquiry-based curriculum and pedagogies with an emphasis on using evidence to critique and construct warranted historical accounts.

Perhaps most powerfully, Allen and Hicks provide specific guidelines and concrete examples for how history teachers, departments, or curriculum producers can create custom LLM interfaces (aka custom GPTs) designed to foster critical and thoughtful historical learning and understanding. Given that history education does not have the billions of dollars of venture capital investment that health technologies or legal technologies companies have to create custom AI models and interfaces, they illustrate how GenAI models can be utilized to create similar custom use models for history education. Instead of the common use of AI to summarize historical readings or generate an essay, these custom LLMs instead can scaffold student analysis of primary sources and encourage and support the development of second order conceptual models of historical causation or continuity. In this way, AI supports students' critical thinking in history instead of replacing it. These models can be used by teachers to support making history more nuanced instead of the simplification of history that occurs with GenAI LLMs that are not customized.

In this way, Allen and Hicks are highlighting the affordances of AI as a supporting medium for engaging students in thoughtful historical inquiry while recognizing the limits of what AI can do if we actually want young people to engage in interpretation and critical historiography and develop an understanding of how media platforms and technology shape historical accounts and representations online. They show how teachers can utilize AI to reinforce the kinds of critical and disciplinary thinking we hope for instead of what most education focused AI platforms are designed to do (e.g., collect data, reduce teaching to transmission).

In short, this book is a primer for history teachers who are technocurious, technoskeptical, or have not even really thought about AI yet—or are afraid to start to. It will help them understand what it is, its limits and concerning bits, and how to start exploring the ways it could help their students reach the goals they have for learning history beyond rote memorization of facts and dates or adoption of historical metanarratives. The book is applicable to a range of courses and grade levels and designed to have a continued life through the near future of AI evolution, especially as the authors plan to update the online companion resources as needed to keep up with shifts in AI capabilities or issues.

Teaching History with Chatty Geeps provides readers with a starting tool kit to be creative in their teaching to address key needs in history education—namely supporting student development in critical analysis skills for engaging in history and for understanding how history may be represented into the future. In sum, the authors have appeared to take some advice from my Patton chatbot: "Don't tell people how to do things, tell them what to do and let them surprise you with their results" (not sure if this is real Patton or movie Patton).

References

Benjamin, R. (2019). *Race after technology: Abolitionist tools for the new Jim code.* Polity.

Cuban, L. (1986). *Teachers and machines: The classroom use of technology since 1920.* Teachers College Press.

PRE-PREFACE

Taking Stock

Before you dive into the book, we invite you to pause for a moment and capture your current thoughts about GenAI and history education. (Hopefully you haven't already skipped ahead, but it's never too late!)

We know that you are not coming to this book empty handed or empty headed. You are here with experiences of being a student, teacher, teacher educator, or a friend/relation of Amy Allen; you have commitments to and interest in history and history education; and you grapple with views, worries, and questions about GenAI and its potential presence in your personal and professional life. You may be skeptical. You may be frustrated, concerned, angry, hopeful, curious, puzzled, or some shifting combination of all of these depending on the day, the context, and your experiences with students, colleagues, institutions, or the tools themselves. You may also have chosen not to use AI at all, and for principled reasons. You are seen:— We want to acknowledge that from the outset. GenAI has opened up a real problem space for educators. It has raised questions about knowledge, authority, authorship, creativity, labor, teaching, learning, and the future of disciplinary work, alongside myriad other legitimate concerns about accessibility, business models, and negative impacts on the environment. Readers will come to this book for different reasons, and they will take different things from it. That is to be expected.

In our work in history education, we aim to prepare students to work with sources in ways that extend beyond basic comprehension. This work involves teaching students to move through the layers of inference and nurturing students to ask where a source comes from, who created it, when and why it was produced, and whose perspectives it represents or omits. Students situate the source in time and place, and they consider how useful it might be as evidence within an inquiry.

We are asking you to do something similar here as a reader.

This book is not a neutral object that you encounter without context. You bring to it your own experiences, commitments, and perspectives that will shape what you notice, question, ignore, accept, or resist. Before you move on to the next section, we invite you to take stock of where you are now so that you can later reflect on how your thinking has been influenced, reified, challenged, or extended.

Use the blank page following this section to record your current thoughts and feelings about Gen AI in relation to teaching and learning in and beyond history education. No one will read this but you, so be as honest as you like. You don't even have to mind your p's and q's. Go for it. Write freely—verbalize your initial ideas as you start this book so that you can return to them later and notice what has shifted.

You may not need any prompts, but if you do, here are some that might get you started:

- Why did I even pick up this book? (Were you told to? Or did you come of your own free will, like Sergeant Howie in *The Wicker Man*?[1])
- What do I hope to get out of this book?
- What do I not want from the book?
- What are my immediate reactions to the emergence/presence of GenAI in the many spheres of my personal and professional life? Why?
- What are my immediate reactions to GenAI in history education? Why?
- What concerns feel most pressing to me?
- What do I worry could be lost?
- What possibilities, if any, do I see?
- Where have my views come from (experience, colleagues, media, policy, students, or something else)?

We also know that all readers do not encounter texts in the same way. You may find yourself agreeing with much of what follows. You may resist many of the claims. You may move back and forth between agreement, skepticism, critique, and curiosity. All of those are valid ways of reading. (If you agree with the text, then Amy Allen says to let her know. If you disagree or have questions, Amy says to let David Hicks know.)

In the Margins

As you move through the book, we hope you will use the margins and spaces around the text as working spaces. You might:

- Jot down questions or reactions
- Highlight claims you want to stress test
- Note where examples and ideas resonate or fall short
- Identify places where you might borrow, poach, adapt, extend, or reject ideas
- Write down what frustrates/annoys you and why
- Record what opens up possibilities and why

1. Note: The 1973 version of the film with Edward Woodward is the best... and his free will got him into a right mess...

Looking Back

Although we note in the introduction that his book does not need to be read linearly, there is value in looking back and reflecting on what you have read. When you reach the end of the book, you will find more blank space for reflection or doodling—at no expense to you! We hope you will consider using this space to revisit your opening notes or doodles and consider how your thinking about pedagogy, disciplinary thinking, history education, random pub facts, *Alice in Wonderland*, or Gen AI has shifted, held steady, or deepened...

If you have not started writing and doodling already, you might ask:

- How did I read this book?
- Was I receptive? Where? (e.g., "I now have posters of the authors on my bedroom wall!" Note: Free pullout poster on page 87.)
- Was I resistant/angry? Where? (e.g., "I now have made effigies of the authors..." Note: Cutout effigies available on page 160.)
- What ideas/perspectives/concerns have stayed the same? Why?
- What has changed? Why?
- What do I now see as possible, necessary, problematic, or still a nightmare? Why?
- What might I borrow, adapt, poach, steal, or experiment with in my own teaching?
- What has shaped my responses as I have moved through the book?

We hope these open spaces help make your own movement and processing visible and legible.

PREFACE: *IN MEDIAS RES*

In 2008, writer and journalist Nicholas Carr asked whether Google was making us stupid. He worried that the web, with its endless hyperlinks and dopamine rewards, was training us to jump from article to article as quickly as possible, barely brushing the surface most of the time, rarely actually reading a full article. He found that the internet wasn't just delivering information but was rewiring our attention spans.

Seventeen years later, journalist Sophie McBain (2025) stood in the MIT Media Lab and wondered whether artificial intelligence had gone one step further. In studying brain-wave research, she noted that writers who composed with ChatGPT showed lower cognitive activity than those who wrote unaided. Regardless of what users felt was happening in their heads, the scans showed that little was stirring. She called this the dawn of a "stupidogenic society," which she went on to describe as a world engineered for frictionless ease, where thinking is merely an optional extra.

In the same year, Torrey Trust and Robert Maloy (2025) joined the conversation, adding their perspective from the field of education. In their study of AI-generated civics lessons, they found that most of what the machines produced was exactly what teachers have spent decades trying to outgrow. Lessons included step-by-step recipes for recall without considering inquiry, context, or diversity. Ninety percent of exercises rewarded memorization; only 6 percent acknowledged marginalized voices. It was as if the ghosts of 1950s civics had been summoned by an algorithm trained on the internet.

And yet, despite these warnings, generative AI (GenAI) is already in classrooms. In a study conducted across four weeks in March and April of 2025, Gallup found that three out of ten teachers now use GenAI tools weekly, saving on average six weeks of planning time each year (Ash, 2025). The promise of efficiency is powerful in a profession often defined by burnout and exhaustion. But efficiency is not the same as enrichment, and we argue that what matters is not how much time teachers save but how they spend it. If GenAI simply accelerates a system that already privileges wide coverage over depth, then teachers will have simply traded one form of overload for another.

Some teachers may choose to reject the personal and professional use of GenAI, and we acknowledge there are benefits to that decision. But even in classrooms where teachers do not use GenAI, there will be students who do (Weingarten, 2025). A 2025 College Board Report found 84% of high school students reported using GenAI tools (Adair et al., 2025). When teachers choose to reject the technology outright, their lack of familiarity leaves them less able to discern when and how students are using GenAI tools, and they also lose the chance to help students learn to use it responsibly, developing critical AI literacy along the way. In our minds, the solution isn't to reject GenAI; it's to rethink the role it can play in the classroom. What would it mean to use those recovered hours not for more grading or content coverage but for reflection, inquiry, and the kind of creative lesson design that GenAI models can't imitate? What would it mean for students to be able to use GenAI not to *replace* learning but to *deepen* their learning?

This is where one of the underlying principles of this book, technocuriosity, becomes important. The technocurious doesn't ask "How can GenAI save me time?" but rather "How can I use this time to think differently?" Carr, McBain, Trust, and Maloy each describe a version of the same concern: as our tools grow more capable, we risk becoming lazy and complacent. The danger isn't that machines will become conscious. The real danger is that we ourselves will become unconscious, content to let predictive text stand in for thought and prewritten lesson plans stand in for pedagogy.

But history offers counter-lessons. Socrates mistrusted writing, and Renaissance humanists feared the printing press (Postman, 1985). Each new medium has disrupted older habits of mind and has also created new possibilities for intelligence.[1] The question then is not whether technology changes us (it always does) but whether we can change with it, deliberately and carefully.

That spirit is present throughout this book. Technocuriosity is a stance of disciplined exploration, based on a willingness to investigate what GenAI can do from the inside—*in medias res*—matched by an insistence on asking what it cannot do (Allen & Hicks, 2025). It's a stance especially suited to history educators, whose craft depends on evidence, context, and interpretation. A technocurious teacher might ask a chatbot to help analyze a historical source, but only after teaching the skills required for source analysis or questioning where the model's version of history comes from, whose stories it omits, and what it smooths over in the process.

How We Wrote This Book (and Why Chatty Geeps Deserves Some Credit)

This book was written with GenAI, but not by it. From the start, our own ideas, frameworks, and classroom explorations shaped its direction. We drafted the core arguments and examples and conducted our own speculative research, drawing on years of teaching and collaborating with students and teachers in real classrooms.

1. We recognize that the environmental impact of GenAI far exceeds that of earlier media revolutions. Training and operating large language models consumes vast energy and water resources, and these costs are not evenly distributed. Still, it's worth remembering that earlier technologies also reshaped ecological systems in their own ways: deforestation for paper, industrial emissions from ink and presses, and the material infrastructures of digital computing. Every medium leaves a trace; what differs now is the scale and speed of extraction, discussed at length in Chapter 12.

At certain stages, we invited ChatGPT (which we affectionately call "Chatty Geeps") into the writing room as a thinking partner. Ethan Mollick's (2024) work on co-intelligence helped us imagine GenAI not as a threat but as a collaborator. His idea that we can "think with" GenAI rather than "through" GenAI impacted our earliest interactions with ChatGPT, and his publications about how to write with GenAI deeply influenced how we approached the writing process (Mollick, 2023a).[2]

We used Chatty Geeps to brainstorm, outline sections, and test the flow of ideas. It occasionally suggested transitions or metaphors that helped us see our own writing more clearly. But every argument, framework, and activity here comes from our disciplinary grounding in history education. Using Chatty Geeps this way pushed us to think harder, not lazier. Sometimes it surprised us with an idea we never would have considered; other times it got things completely wrong, flattening nuance or missing the point. Those jagged edges of the frontier (Mollick, 2023b) were actually some of the moments where we learned the most. They reminded us why human judgment, disciplinary grounding, and a sense of humor all matter when working with GenAI. To be clear: ChatGPT didn't write this book. It couldn't have. But it did help *us* write this book by helping us think.

For example, when we were designing the Classroom Lifts (exercises throughout the book categorized as Low Lift, Mid-Level, and Ambitious), we'd often ask Chatty Geeps to generate a few options just to see what might spark. We'd try them out, test them against classroom experience, keep a few, and toss most. The same thing happened with our reflective questions. In these cases, we used GenAI as a jumping-off point but never as the finish line.

Our outlining process was also recursive. Before writing each chapter, we fed identical summaries of what we wanted to cover into our personal GPTs. The responses we got back were sometimes helpful, sometimes conflicting, occasionally ridiculous, yet somehow, always flattering. But! These responses forced us to clarify what we actually wanted to say:—[3] from those exchanges came the giant sheets of chart paper that became the skeletons of every chapter.

---

2. At the time of writing this book, we haven't seen many people acknowledge if/how they are using GenAI in their writing. As you will see in Chapter 12, we require our students to complete an AI Awareness Statement. This section of the text is our AI awareness statement, acknowledging when and how we used GenAI as part of our writing process.

3. We have traditionally used em dashes in our writing; however, the rise of GenAI has led to a rise in the use of em dashes, and often, people associate their use with writing that was produced by a chatbot like Chatty Geeps. Even as we acknowledge the way we used Chatty Geeps in our writing, we must also acknowledge that it has impacted our writing styles; now, we avoid em dashes as much as possible. In lieu of these, David is trying to bring back "dogs bollocks (https://en.wikipedia.org/wiki/Compound_point)" punctuation, and you will see we intentionally use these instead throughout this text:— It was only during copy editing that David learned that for all his enthusiasm for the dog's bollocks :—, he had been whipping it out rather too freely for almost any punctuation occasion and, worse, had been consistently using the lesser :–. Having now looked at them side by side, David sees that size does, apparently, matter. While the full em dash version of the dog's bollocks has a certain flair, it also strikes him as somewhat overdramatic. He remains partial to the more restrained aesthetic, however (mis)used.

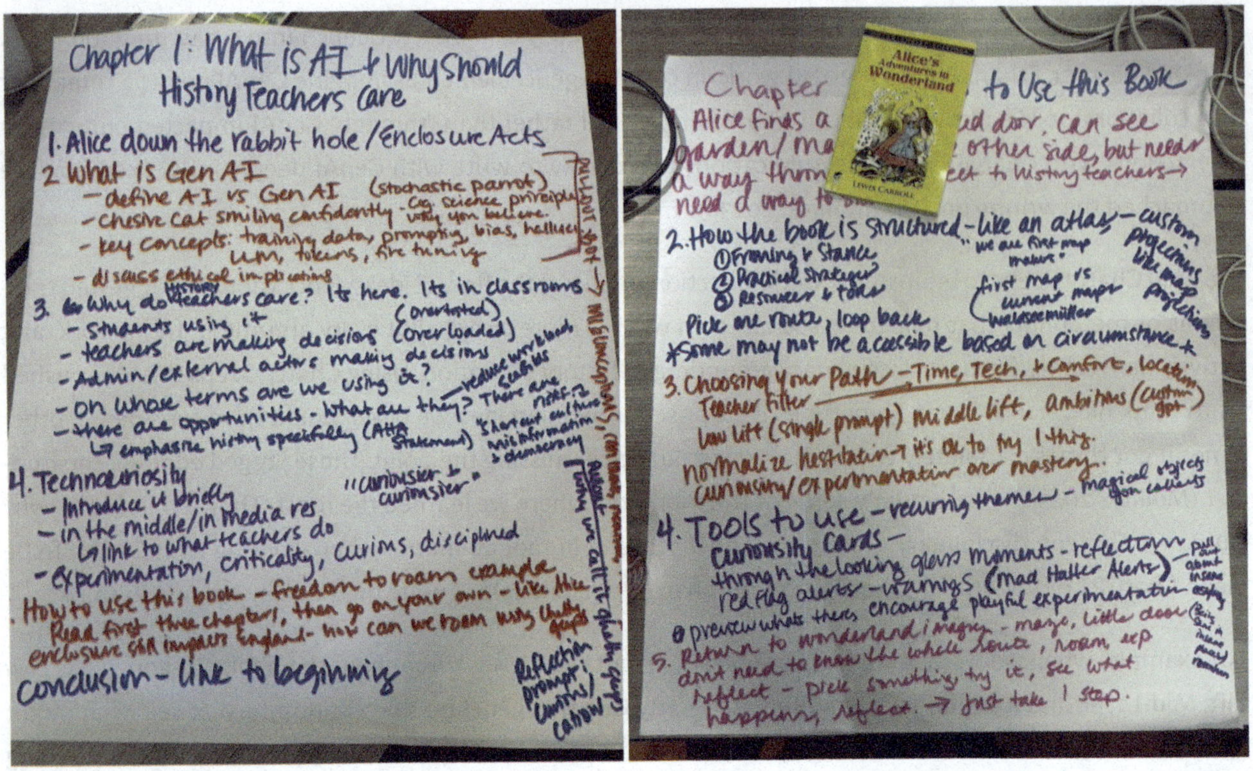

Figure P.1: Outlining process

Even the book's literary framing based on *Alice in Wonderland* began as a conversation with Chatty Geeps. Early in the process, Amy (the bossy co-author) asked Chatty Geeps if there was a story we could use to frame the book, similar to how we used *Seven Blind Mice* in the CITE editorial that introduced our technocurious framework (Allen & Hicks, 2025). One of the twenty vague suggestions was *Alice in Wonderland*. That prompt triggered a memory of a fantasy literature workshop Amy attended at Oxford led by Fran Kohlt (https://franziskakohlt.com), whose brilliance we deeply admire. The literary examples that run through this book trace back to that workshop or were developed based on the information Amy learned there; they are not the result of anything Chatty Geeps generated later. In fact, anytime we tried to get Chatty Geeps to brainstorm potential Alice connections, they were shallow and repetitive.

Our long partnership with the Library of Congress and the Teaching with Primary Sources (TPS) program (https://www.loc.gov/programs/teachers/about-this-program/teaching-with-primary-sources-partner-program) also shaped how we wrote this book. For years, we have worked with teachers to use historical sources as lenses for disciplinary thinking, including consideration of how meaning, power, and evidence intersect in the making of history. That work is woven into every chapter here. Sometimes we used sources we already knew from our classrooms or research. Other times we asked Chatty Geeps to help us locate new sources, treating it as a kind of digital research assistant. Although it was not always successful at identifying real sources, it often gave us ideas on where to start looking. Our experiences with the Library of Congress helped us see GenAI as a new kind of archive: not a replacement for historical evidence but a tool that can help teachers find resources. As we shared in an NCHE presentation in early 2025, GenAI also allows teachers to rethink what searching looks like, using Chatty Geeps as a partner to correct or overcome OCR issues in historical sources held in databases such as *Chronicling America* (https://www.loc.gov/collections/chronicling-america/about-this-collection).

These examples demonstrate the way our process mirrored the stance we advocate throughout this book:—technocuriosity in practice. We experimented, questioned, revised, and occasionally disagreed with Chatty Geeps and with each other. The work you're holding now is the result of that ongoing, recursive conversation.

That work is also the result of other technological advancements over the past several decades. Throughout the process, we relied on Grammarly suggestions to ensure we didn't accidentally write run-on sentences (David) and Microsoft Word's spellcheck to make sure we spelled everything correctly (Amy). We used Google as a thesaurus and dictionary to make us sound smarter than we probably are, and we used database search engines like ERIC to download online sources digitally rather than the physical card catalogs and photocopiers we both remember from our years as college students. Amy sometimes used the talk-to-text feature on her iPhone, then copied and pasted what she had orally dictated into her phone as a first draft of the case studies included. All of these tools are part of a trajectory, and we recount them here to demonstrate the ways in which writing has been shifting for decades. Studies show that writing by hand, typing, word processing, and using AI-assisted writing tools each change how we compose, revise, and think on the page (Hayles, 2002; Kirschenbaum, 2016). The result? The act of writing no longer just happens in our heads. Instead, it happens with our tools. So when we say we wrote "with Chatty Geeps," we are acknowledging both continuity and change. This book is part of a longer history of humans writing alongside machines. In that sense, this book is doubly historical, simultaneously addressing the discipline of history and serving as a historical artifact of writing practice.

This way of thinking about the writing process also prompted us to consider the way we were thinking about reading in a digital age. To that end, this book is available open access online and is intentionally interactive, with the ability to read laterally and engage in collaborative notetaking through Hypothesis.

We owe a special debt to Celia Lury (https://warwick.ac.uk/fac/cross_fac/cim/people/celia-lury), whose work David (the chronically late co-author) loves (something which he will tell anyone who stands still long enough).[4] Her essay "Going Live: Toward an Amphibious Sociology" (2012) gave us a vocabulary for how it feels to work from within a living, shifting system rather than studying it from afar. After David sent the article to Amy, she read it multiple times on a plane ride home from Disney World (because of course she did). Both of

4. She may not remember him, but David was a student in Ceila's Women's Studies course at Lancaster University in [year redacted].

us immediately saw its connection to what we were trying to make sense of. That was the spark that eventually ignited the technocurious framework.

So no, Chatty Geeps didn't invent our ideas or physically write this book.[5] But it did remind us, provoke us, and help us hold a mirror to our own thinking. In the best moments, it was like having a very patient teaching assistant who almost always quoted the wrong book but also always got us talking again.

On All These Victorians

Before we move on, a brief acknowledgment (and perhaps a small apology). You'll notice that many of our examples originate from Victorian England: botanical gardens, mental asylums, spirit photography, the moralizing of insects, and of course, Alice's Adventures in Wonderland. *That is not an accident. We chose to situate this text there because Carroll's world of curiosity and nonsense gave us a historical mirror for our own moment of technological bewilderment. But we also recognize that Victorian England was a deeply racialized, imperial space. Its metaphors of cultivation, improvement, and order were built atop exclusion, hierarchy, and empire. The same world that produced* Alice in Wonderland *also produced eugenics, enclosure, and the global museum.*

We use these scenes not to celebrate the white, imperial imagination of the nineteenth century but to interrogate it by asking how systems of order and value take hold and by investigating how they persist in GenAI algorithms and educational assessments. In the appendices, we've included Curiosity Cards that pair many of these historical source examples with short activities teachers can use to introduce the same critical lens in their own classrooms. Each card draws on a historical source (including many from the Victorian era) and invites students to ask questions like:

- *Who made this system of knowledge?*
- *Who was included or excluded?*
- *What counts as "sense," and who gets to decide?*

In other words, the cards do what we hope each part of this book does: use history to question how we know what we know. Our hope is that readers will see those examples as both historical and diagnostic, unearthing ways of thinking with the past to understand the strange, unequal systems of the present.

And if at any point you find yourself thinking, "Amy might be a little too into Alice in Wonderland,*" you're right. She is. We've accepted it. It's fine. Everything is fine.*

5. In the interest of full disclosure, if you enter this book into AI-detecting software, it will say it's about 20 percent AI generated. We like to think that is due to Amy's exceptionally good grammar. ... And actually, we're a little disappointed it isn't higher!

Why You Might Want to Read This Book

This is a book in motion because the landscape it describes keeps changing beneath our feet. Any book about GenAI will be out of date the minute it's published, and we're okay with that. What matters isn't keeping pace with the updates but learning how to think alongside them, and we hope that this book inspires you to do your own exploration and take these ideas further than we could in book form.

There are many ways you might find your way in this book:

- If you're a history educator curious about GenAI but unsure where to start:— we have ideas.
- If you're AI-curious but not sure what it has to do with history:— we have ideas.
- If you're interested in a case study of what it looks like to play with GenAI:— we have ideas.
- If you don't care much about GenAI but enjoy historical asides and primary sources and want to maximize your chances at the Tuesday pub quiz:— read on.
- If you just want to know what's going on in Amy's head and how on earth *Alice in Wonderland* connects to a framework called technocuriosity:— read on.
- And if you're secretly hoping to decode hidden messages about people we don't particularly like:— well, good luck. (You won't succeed, but you'll still learn something.)

Most of all, this book is for teachers who are trying to navigate a world that feels like it's changing too quickly for their lesson plans to keep up. We often say we don't know what we are talking about. We aren't experts on GenAI (who could be?), and we aren't promising that this book will bring mastery. But we hope it can be a starting place. We invite you to wander with us through Wonderland, through Victorian England, and through the unfamiliar terrain of GenAI without losing sight of who you are as a history educator.

In these chapters, you'll find strategies for using GenAI to design prompts, analyze sources, and scaffold reflection with a basis on ethics and humanity. You'll see examples from real classrooms and professional development workshops, moments of failure (there are lots of those, and we have more), and glimpses of what we think the future might hold.

So take a deep breath, grab your curiosity, and like Alice, step through the looking glass. The world on the other side is strange, but we will attempt to make sense of it together.

1. WHAT IS AI, AND WHY SHOULD HISTORY TEACHERS CARE?

The children's novel *Alice's Adventures in Wonderland* (1865) begins with Alice chasing a white rabbit down a hole, leaving behind the world she knew for one governed by contradictions, nonsense, and ever-changing rules. While this familiar story is often discounted as a silly story for children, contemporary and historical references are woven throughout the text and its 1871 sequel, *Through the Looking-Glass* (e.g., Kohlt & Houyaux, 2024). Charles Dodgson, an Oxford lecturer better known by his pen name Lewis Carroll, wrote both books during the height of the British Enclosure Acts. Between 1604 and 1914, Parliament passed more than five thousand acts that forcibly enclosed over 6.8 million acres of land across England (Fairlie, 2009). Communal spaces such as meadows, forests, and grazing fields were fenced, sold off, and redrawn on new maps. Local traditions and ways of knowing, intimate relationships to place, and systems of shared use were bulldozed by productivity, regulation, and profit as the rules governing public and private spaces shifted. In Wonderland, the fictional character of Alice, like many of her historical contemporaries, encounters fields and forests neatly squared off like a chessboard amid new and ever-changing rules designed without input from the constituents impacted. She is asked to play a game whose outcomes are fixed in advance, governed by rules she did not create and cannot question.

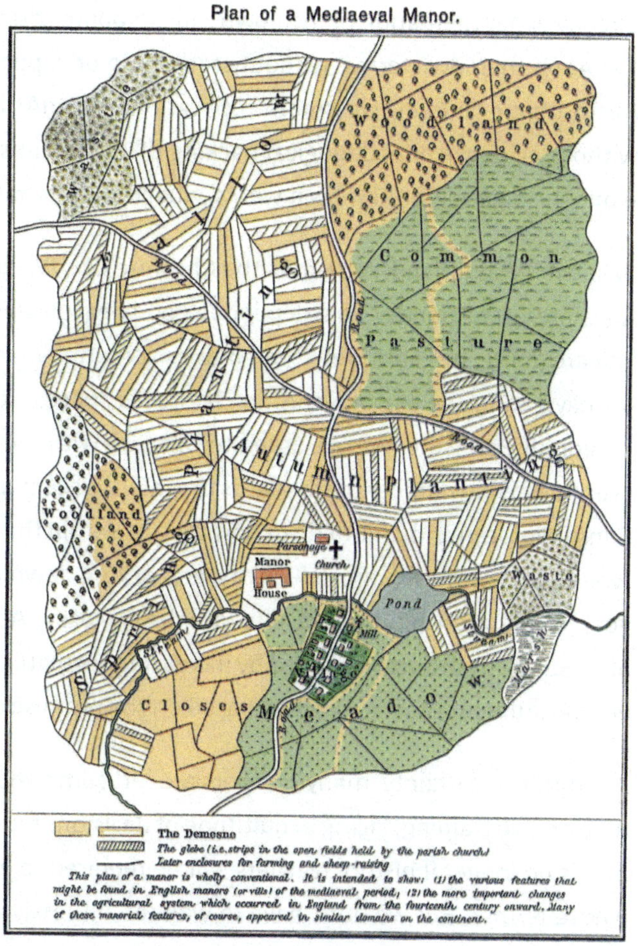

Figure 1.1: Prior to the Enclosure Acts, medieval manors included open fields and common pastures that were open to all. After the Enclosure Acts, many of these spaces were blocked off, establishing clear boundaries and preventing public access.

Similar to the impact of enclosure in England, teachers have spent the last two decades watching their domains shrink as curricular mandates have tightened and instructional time has become increasingly scripted (American Historical Association, 2024; Au, 2007; Goldstein, 2025; van Hover et al., 2016). Navigating this landscape during the rise of generative artificial intelligence (a tool that produces immediate and authoritative responses without providing evidence, context, or comprehension) feels a little like falling down a rabbit hole. In some districts, administrators have begun mandating the use of GenAI, but often without clear expectations on how to implement it (Singer, 2025a). In other districts, teachers have fully banned the use of tools like ChatGPT without any experimentation of their own, sidestepping critical conversations and leaving space for students to use GenAI on their own terms, without scaffolding or support. Increasingly, tech giants like OpenAI, Google, and Microsoft are peddling classroom applications of GenAI, heralding its transformational power in the classroom without evidence or pedagogical backing of such claims (Singer, 2025b). All of this is happening in real time, while the logic of GenAI (how it works, what it does, what it knows) remains shrouded in mystery.

While in Wonderland, Alice asks the Cheshire Cat, "Would you tell me, please, which way I ought to go?" The Cheshire Cat responds, "That depends a good deal on where you want to get to." If you don't know or care where you are going, he continues, "then it doesn't matter which way you go." This sentiment is a spot-on metaphor for navigating the use of GenAI in classroom spaces. Today, as history teachers, we are faced with our own version of enclosure. We are asked to carry on with our work, teaching historical thinking while covering an ever-increasing number of standards that always seem to be changing. Without a clear understanding of what GenAI tools can and cannot do or how they fit within the pedagogical aims of history education, it is easy to get swept up in promises of productivity, embarking down a road that often leads further away from the deeper goals of historical inquiry. And, like with the Enclosure Acts, calls to use and implement GenAI tools in the name of productivity are accompanied by increasing regulations without input from those who are impacted (teachers and students) and are instead based on the promise of exponential profit for the tech industry.

Amid this uncertainty, many teachers are left somewhere in the middle, navigating these new tools with no map and no training, facing a multitude of decisions for which they feel unprepared. At this juncture, the question is no longer if history teachers will be required to engage with GenAI. The questions are "How?" and "On whose terms?" But we don't have to wander aimlessly on this path alone. This book invites history teachers to engage in thoughtful, disciplinary, practice-based conversations about how we can respond as experts at attending to issues of complexity, evidence, and interpretation. This chapter opens the book by laying out the terrain. We begin by unpacking what GenAI is and what it isn't, why it matters for history educators, and how we might navigate this ever-changing landscape by, like Alice, embracing a stance of curiosity alongside experimentation and critical grounding. We invite you to jump down the rabbit hole with us, but unlike Alice and the Cheshire Cat, we're not wandering. We have a map.

What Is Generative AI?

Artificial intelligence (AI) broadly refers to the capacity of machines to simulate human cognitive processes such as learning, reasoning, and decision-making. Many educators are already familiar with forms of AI embedded in everyday technologies. For example, tools like Grammarly offer stylistic suggestions, and online platforms recommend search terms or other information based on previous searches. These tools analyze patterns in existing data and make decisions or predictions based on preprogrammed rules or user behavior.

Generative artificial intelligence (GenAI) is a more advanced and rapidly developing subset of AI. Unlike traditional AI, which is largely reactive and rule-based, GenAI models are designed to generate entirely new content (text, images, audio, or code) based on the patterns they have identified in vast datasets. These models do not retrieve prewritten material or rely on static templates. Instead, they use probabilistic prediction to construct novel outputs, word by word or pixel by pixel, in response to user prompts. Systems like OpenAI's ChatGPT (Chatty Geeps), Google's Gemini, Notebook LM, and Anthropics's Claude are some of the most well-known examples that you may have heard of.

Why We Call It "Chatty Geeps"

By now, you've probably noticed that we don't call ChatGPT by its full name. Throughout this book, we refer to ChatGPT as "Chatty Geeps." Why? Because we want to signal something important: this is not magic, not a person, and not a colleague. It's a spectacular guessing machine that predicts the next likely word based on patterns in its training data (flawed normative data).

So why the silly nickname? Partly for our own amusement:— we poke fun at the tool even as we test its possibilities. Saying Chatty Geeps disarms some of the anxiety surrounding the tool, making space for curiosity and experimentation rather than intimidation. But also, the word "Chatty" captures what these tools actually are: stochastic parrots (Bender et al., 2021). They are endlessly chatty and sometimes persuasive, but that does not necessarily mean they're insightful. The voices Alice encounters in Wonderland are confident, talkative, and sometimes amusing. Similarly, GenAI models are eager conversational partners, fluent and fast, but their conversation is not thoughtful or grounded in evidence. Calling it Chatty Geeps reminds us (and our students) that this is not a historian, teacher, or a truth-teller. It is a program that generates plausible-sounding text based on statistical patterns. To misquote (probably) Sir Charles Dilke, "There are four lies: lies, damn lies, and statistical patterns, and Chatty Geeps" (Menter, 2020).

Renaming the tool also serves as a reminder of teacher agency. Calling it Chatty Geeps reminds us, our students, and our colleagues not to be seduced by the tool's fluency and sexy talk. Just as Alice had to decide which voices to trust in a world full of nonsense, teachers and students can use the nickname to signal that they will not take GenAI's words at face value.

GenAI models are trained on an immense body of text:— billions of words sourced from websites, books, academic articles, public records, and other online content. From this training data, the model learns to identify linguistic patterns, syntactic structures, and semantic relationships. When a user submits a prompt, the model generates a response based on the most statistically probable next word, followed by the next, and the next, and so on. In other words, GenAI is a spectacularly good guesser (even better than the predictive text on your iPhone). The result of its guesses is a coherent, contextually appropriate piece of text that appears, on the surface, to be both knowledgeable and intentional.

However, the training data itself is not neutral or evenly representative. Just as enclosure transformed common lands into private property without consent, GenAI appropriates creative and scholarly work from the commons of the internet, redrawing boundaries of ownership and access. Much of it comes from public internet content, but this includes material that was scraped without permission, such as copyrighted books, academic work, journalistic reporting, and creative writing (Carson, 2025). In many cases, the original creators of this content were neither asked nor compensated. More broadly, the perspectives and patterns the model learns are shaped by the most prominent voices online, further overlooking the absent, misrepresented, or marginalized. These practices raise pressing concerns about intellectual property, plagiarism, representation, and our responsibilities when using tools built on appropriated work.

It is also important to underscore that these systems are not actually reasoning, understanding, or interpreting in any human sense. They do not know historical context, and they do not possess the ability to distinguish between valid and invalid claims. As such, the outputs they generate may sound fluent and persuasive while still being incorrect, misleading, or entirely fabricated. When a GenAI model produces content that has no basis in factual knowledge but presents it with confidence and coherence, this is referred to as "hallucination." Several key terms are useful to understand when engaging with GenAI platforms:

- **Prompting** refers to the process of giving GenAI a task or question. The quality and specificity of the prompt often shapes the quality of the response. In this way, prompting functions less like a query in a search engine and more like a starting point for dialogue.
- **Tokens** are the basic units of language that the model processes. A token might be a word, part of a word, or a punctuation mark, depending on the model. Token limits determine the length and depth of the interactions.
- **Training data** refers to the dataset on which the model was built. The scope and quality of this data also directly influence the kinds of responses the model can generate. It also carries with it the biases, omissions, and blind spots of the original sources.
- **Fine-tuning** is the process by which a base model is adjusted to perform a particular task or adopt a specific tone. This might involve refining a prompt, but it can also refer to the creation of a history-specific model that draws from curated archives or vetted historical sources. Fine-tuning adjusts the model's

internal parameters, resulting in more consistent and aligned responses.

- **Hallucination**, as noted above, describes instances where GenAI "invents" information that does not exist in the training data. In educational settings, this can lead to false historical claims, fabricated citations, or oversimplified interpretations of complex events.

For history educators accustomed to working with vetted sources and clear evidence, the idea that a tool can hallucinate its way through a historical explanation is dangerous. The outputs generated by GenAI are often plausible, but plausibility is not the same as truth. As a result, the use of these tools in educational contexts requires a critical awareness of both their affordances and their limitations.

The parallel between the arrival of GenAI in schools and the historical process of enclosure continues to resonate here. In many ways, GenAI is a tool that reshapes what is visible, what is knowable, and what counts as credible. Just as the enclosure of common lands redrew physical and social landscapes in England's past, the widespread use of GenAI is redrawing epistemic boundaries, shifting how information is produced, who produces it, and how it circulates within the classroom. To engage with GenAI as a history educator, then, is more than just adopting a new tool. True engagement means navigating a new kind of terrain where meaning is generated by probability and the appearance of authority can mask understanding. The first step toward engaging with GenAI intentionally involves understanding what it is and what it is not.

Why Should History Teachers Care?

We are not unaware of what is happening in schools:— we both have children in high school. This academic year, they both have brought home forms outlining expectations of what responsible GenAI use looks like. In some cases, teachers have enacted full bans, asking parents to sign forms promising that their students won't use it.

In other cases, teachers are asking students to complete physical, in-class writing assignments to get to know their writing styles, which helps teachers recognize the use of tools like Chatty Geeps. And through it all, both of our children acknowledge that students are still figuring out how to use GenAI tools to complete their assignments; however, due to these rules, they are doing so without the benefit of a teacher who can scaffold the process and help students learn about critical AI literacy.

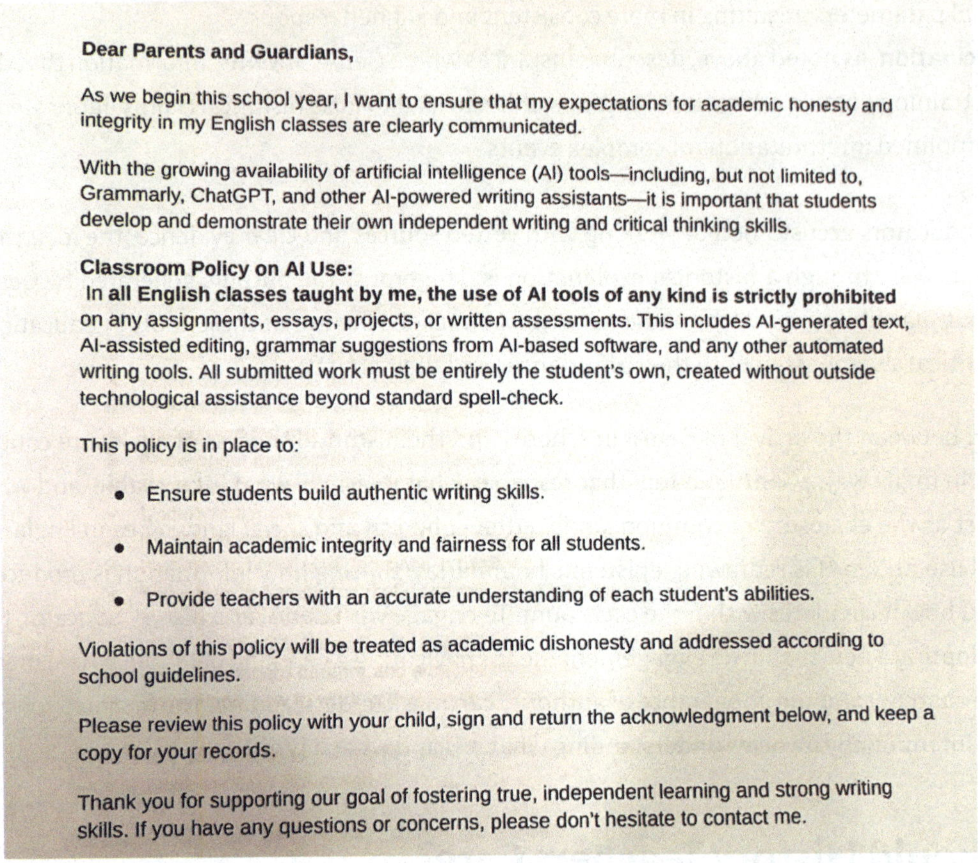

Dear Parents and Guardians,

As we begin this school year, I want to ensure that my expectations for academic honesty and integrity in my English classes are clearly communicated.

With the growing availability of artificial intelligence (AI) tools—including, but not limited to, Grammarly, ChatGPT, and other AI-powered writing assistants—it is important that students develop and demonstrate their own independent writing and critical thinking skills.

Classroom Policy on AI Use:
In **all English classes taught by me, the use of AI tools of any kind is strictly prohibited** on any assignments, essays, projects, or written assessments. This includes AI-generated text, AI-assisted editing, grammar suggestions from AI-based software, and any other automated writing tools. All submitted work must be entirely the student's own, created without outside technological assistance beyond standard spell-check.

This policy is in place to:

- Ensure students build authentic writing skills.
- Maintain academic integrity and fairness for all students.
- Provide teachers with an accurate understanding of each student's abilities.

Violations of this policy will be treated as academic dishonesty and addressed according to school guidelines.

Please review this policy with your child, sign and return the acknowledgment below, and keep a copy for your records.

Thank you for supporting our goal of fostering true, independent learning and strong writing skills. If you have any questions or concerns, please don't hesitate to contact me.

Figure 1.2: Letter shared with parents about classroom policy on AI use

In the same school, down the hall, there is a history teacher who is really trying to play with GenAI and thinking about the implications of what tools like this mean for history pedagogy. He attended a workshop we held in the summer, developed his own rules for when, why, and how students will be asked to engage with GenAI, and has begun experimenting with building custom GPTs that will provide feedback for students completing document-based questions (DBQs[1]).

The fact that these cases can both exist in the same school demonstrates that some administrators are facing the rise of GenAI tools by simply ignoring them and allowing teachers to set their own parameters for how teachers and students will engage. But in other districts, administrators are responding more actively, purchasing school-wide subscriptions to programs like School.ai, Course Companion, or Magic School and rolling them out with or without professional development to help teachers understand how these models work (Singer, 2025a).

1. More on DBQs in Chapter 3!

Most of the programs purchased by school districts are "wrapped" versions that come with guardrails and branding but have little connection to the actual disciplinary aims of history teaching. These programs advertise that they are tied to standards, but the developers behind the programs are unknown. The professional development that accompanies the implementation of these programs is also unlikely to center on the disciplinary aims of history teaching and, perhaps more problematically, is often developed and sold by the creators of GenAI platforms like OpenAI or Google—stakeholders unlikely to prioritize the pedagogical goals of history educators. Without history teachers' voices in the conversation, the rush to use GenAI has moved toward implementation that positions platforms like Chatty Geeps as tools for efficiency rather than as resources to support learning history as an inquiry-based discipline.

History teachers bring a unique perspective to the table. The American Historical Association (https://www.historians.org) (AHA) and the National Council for the Social Studies (https://www.socialstudies.org) (NCSS) remind us that history and the social sciences are not simply about teaching kids to memorize random facts. In its ideal and best form, history teaching is about inquiry, interpretation, inferences, and making evidence-based claims. These habits of mind (https://ncheteach.org/resources/historys-habits-of-mind) are exactly what teachers need to cultivate in students to help them read the world:— past, present, and future. While GenAI may enable students to "cheat" by generating a perfect five-paragraph essay on a historical topic, it may also enable students to engage in the sort of meaningful inquiry that we suggest is at the heart of good history instruction.

We believe history educators are uniquely positioned to approach GenAI technology with disciplinary care and ethical grounding. It is with that mindset that we present this book and offer recommendations for working with Chatty Geeps. Rather than wrapped platforms like School.ai, we chose Chatty Geeps as the basis for this book because of the level of freedom and flexibility it provides to speculate, experiment, probe, and stress-test a GenAI that is not already externally controlled or narrowed by external sources.[2] Using a tool with the level of freedom Chatty Geeps provides gives an extra degree of agency to educators, allowing them to critically expand their pedagogical imagination rather than reducing GenAI to a productivity app geared toward the development of lesson plans. Restricting the use of GenAI platforms to such shallow tasks reduces the ability of teachers to be creative and short-circuits opportunities for facilitating deep learning with students through scaffolding disciplinary thinking and ambitious history teaching.

---

2. As we complete the final copy edits of this book in March 2026, we are mindful that the ethical questions surrounding GenAI reach far beyond classroom use and involve the companies building these tools and the business models shaping their development. OpenAI's recent defense agreements, including work with the U.S. Department of Defense, remind us that these systems are tied to larger questions of power, profit, and public accountability. With open eyes, we have chosen to continue to use Chatty Geeps as we write about it while, at the same time, acknowledging that the tool is not neutral or beyond criticism and the industry behind it is not ethically settled. As we continue to engage with Chatty Geeps, we have also begun to play with and stress test other tools, including our university's own LLM, Claude, Gemini, Notebook LM, Stanford's STORM, and Perplexity. For us, critical engagement with these tools is a necessity because of the ways in which they are shaping the conditions in which teachers and students work.

That is not to say that productivity applications are bad! Productivity supports matter when teachers are expected to do so much and when the stakes are human. In our workshops, teachers talk about exhaustion due to pressure to grade student work on weekends, differentiate lessons for students with wildly different needs, and design supports for multilingual learning. We believe GenAI can and should lighten that load for teachers, and we touch briefly on instructional design in Chapter 6. However, that is not the purpose of this book. Instead, we want to focus on how Chatty Geeps can help educators think deeply about what it means to learn. We want to position this tool as a way to support learning in a discipline we love. We want to learn which doors GenAI can open for students and to consider how Chatty Geeps can help expand opportunities for students to harness their curiosity, sharpen their historical questions, refine their evidence-based arguments, discuss issues of significance, consider the ethics of representation, fundamentally make claims that are based on evidentiary warrant, and engage in self-regulated learning.

Many reading this book may feel the risks of using GenAI in this way outweigh the benefits. It is true that there is potential for GenAI to become a shortcut around thinking and doing history. If history instruction is grounded in the passive reception of knowledge, tested through multiple-choice quizzes and standard essays, then this concern surrounding GenAI is well founded. In these cases, tools like Chatty Geeps will hold up a mirror to the shallow assessment practices present in some history classrooms and reveal the inadequacies of such methods. Because of the unique opportunities afforded by GenAI, it has the potential to be used in ways that position the tool as a powerful scaffold to support a deeper and flexible learning of history.

Finally, we recognize that the fluency of GenAI outputs can be both alluring and deceiving. GenAI generates text that is grammatically polished, confident in tone, and just sounds sexy, alluring, or appealing. The issue is style over substance. The path of least resistance for student learning is to copy and paste. This temptation has always been there. Many of us will remember copying chunks of information from textbooks or Wikipedia when we were students. GenAI slop and automated plagiarism are extensions of the same type of plagiarism that has existed for years, though it is now easier to get away with it. However, even if a paragraph on the American Revolution generated by GenAI appears convincing, it can very well contain significant errors. This is a danger and a gift. If students take the information at face value, they are set up for failure. This common mistake is why blanket policies that prohibit the use of GenAI are problematic. If schools are where students are meant to learn how to navigate the world, they should also be spaces where they learn how to use technology tools with intention and nuance. The use of GenAI transforms into a gift if the teacher is able to frame its use carefully, modeling how to critically unpack its output. Organizations like the AHA support this type of use. They caution against blanket bans and instead ask history educators to teach students how to approach and interpret GenAI outputs (AHA, 2025). The output becomes something to corroborate, contextualize, and critique. By building critical AI literacy in this way, students learn to treat GenAI outputs as they should any historical or contemporary source.

In our opinion, the question is no longer whether history teachers should engage with GenAI. Students already are, and we are already a step behind them. The question is now how and on whose terms do we engage with GenAI as educators? If history teachers do not engage but prohibit, hide, step back, and simply critique, then others who know nothing about our discipline will shape the development of GenAI tools, emphasizing perceived productivity over learning and minimizing its potential in the history classroom. The terms are and will continue to be set by vendors, entrepreneurs, and administrators. Students will continue to experiment and play with the tools in isolation. History teachers will lose the opportunity to shape the use of GenAI with disciplinary care, expertise, and ethical grounding. This book is our effort to step into that space and help shape the development of the intersection of GenAI and history education.

Technocuriosity

One way we find ourselves contributing to this space is through the development of a stance we call technocuriosity. As Alice travels through Wonderland, she often repeats the phrase "curiouser and curiouser" as the world bends logic and defies her expectations. Her comment is a reflection of the odd places and spaces that make up the imaginary world in which she finds herself. In contrast, when we talk about curiosity as it relates to this stance, we are referring to the ways we find ourselves engaging with the imaginary world created by GenAI. The more we play with and engage with these tools, the more curious we become about their capabilities and their applications for education.

In our work with teachers and across the research about technology in education, we have seen three dominant orientations emerge. On one side is technoskepticism, the position that GenAI is too risky, biased, or ill-suited for education to be used at all (Pleasants et al., 2023; 2024). This stance is important for its ethical perspective, reminding us, as we have outlined above, that tools like Chatty Geeps are not neutral and are imbued with structural inequalities. On the other side is technoromanticism, the enthusiastic adoption of GenAI as a transformational tool that has the potential to revolutionize learning (Fitzpatrick, 2024; Watson et al., 2024). This stance is often generative in its willingness to experiment with emerging tools and imagine new pedagogical possibilities, but it does not always fully consider the disciplinary and ethical complexities associated with learning.

We find ourselves adopting a third, alternative stance. We identify as technocurious, a perspective that embraces a certain degree of fluidity, defined by its constant movement on a spectrum between two poles. Technocuriosity occupies a messy middle ground. It resists fast adoption but also blanket refusal. It invites teachers to approach GenAI not as a simple solution but as an unfolding problem space (Lury, 2021). To be technocurious is to work *in medias res*, from the inside, using experimentation as a way to think about what this developing technology might mean for teaching and learning history (Allen & Hicks, 2025; 2026).

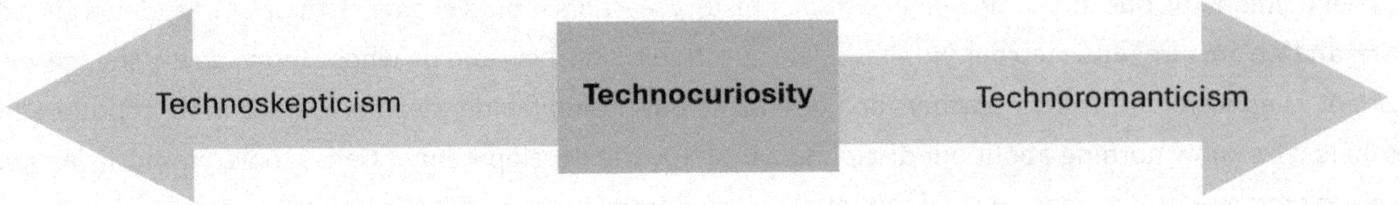

Figure 1.3: Locating technocuriosity on a spectrum

Several key tenets make up this stance. **Situated liveness** asks us to engage with GenAI in the messy present, not just in theory. To make sense of it, we must use it, test it, and respond to it. How can we, as educators, respond to students' use of technology if we are not also using the technology and attempting to make sense of it? **Speculative ethics** asks not just "Does it work?" but "What futures are we building when we use this?" This tenet responds to the technoskeptics who call for critical reflection on our use of GenAI. In line with this idea, you will find space throughout this book to engage in critical reflection as you consider potential applications in your own educational settings. **Relational configuration** positions GenAI as entangled in schools, systems, and communities. As evidenced by the experiences of our own children, GenAI does not exist in isolation. It exists in classrooms, where teachers must navigate student perceptions, follow district guidance, and adapt to varying levels of technology resources. Across these tenets, technocuriosity engages in **figural thinking**, treating GenAI not only as a tool but also as a figure carrying symbolic weight about knowledge and the future. This leads to both **methodological invention**, a willingness to improvise, hybridize, or invent new ways of teaching and researching, and **reflexive co-creation**. Educators and students are actually co-constructors of GenAI's role in learning.

Adopting a technocurious stance helps us view the current moment not as a crisis or cause for moral panic but as an opportunity. It provides a chance to consider what we mean by learning history and to shape pedagogies and experiences for teachers and students. Adhering to these principles does not guarantee safe passage, but it generates opportunities and spaces for cautious exploration, ethical reflection, and creative risk-taking within the discipline of history education.

Freedom to Roam

We began this chapter with the Enclosure Acts, when common spaces were fenced off, redrawn, and claimed by others. But the story did not end there. Centuries later, in the year 2000, England passed the Countryside and Rights of Way Act, establishing the freedom to roam on mapped "access land" (Natural England, 2014). This public access campaign provided access and rights of way to lands that had been closed and restored the right to wander across some privately owned land. Although it did not establish a universal right to roam or create new rights of way, it increased lawful public access to certain parts of private land. From this example, we see that boundaries are not fixed forever; rules can be reimagined and spaces can be reclaimed in the public interest.

We see a similar possibility in how educators might approach GenAI. As the technology has advanced, spaces have been enclosed. Teachers have already been given rules on how to engage with GenAI and are often provided with subscriptions for "safe" but limited models. The foundation has been laid for how GenAI will be enclosed in classroom spaces, with the terms of access and use increasingly set by vendors and not educators. But, we argue, teachers are professionals with pedagogical decision-making power that should have the freedom to roam across varied platforms and to make decisions that stay true to the disciplinary goals of history education. We hope this book provides guidance on how to move forward in this vein.

But we also need to be transparent. We share the use of a paid version of Chatty Geeps that allows us to create custom GPTs and work with greater memory and token limits. That is not an endorsement of one product over another. However, it does reflect the choices many educators now face and the complex, contradictory questions about whether and how to use GenAI at all. For us, the value lies in generating opportunities for innovating and experimenting with the tools as they currently exist. We see ourselves less as rule followers on a fixed path and more as mapmakers, charting paths that fellow history educators might also explore, adapt, and redraw.

This book is set up with two parts. The first three chapters provide a foundation for understanding the current (quickly changing) landscape, explaining what GenAI is, the technocurious framework, and how to evaluate it through a disciplinary lens. Then, in the fourth chapter, we provide more guidance on how to navigate the second half of the book, which is modeled after a Choose Your Own Adventure book. Based on your own comfort level, classroom space, and available technology, you'll move around to the chapters that sound the most interesting or align with your teaching goals. Throughout each chapter, we provide practical strategies rooted in an Alice metaphor and examples supported by historical sources, as well as sample prompts, case studies, and potential red flags.

What follows is not written for computer scientists or technology experts. It is written for busy, thoughtful, and sometimes skeptical teachers who understand the complexity of their classrooms better than we do. As you read about these ideas, we want to encourage you to also try them. It's normal to feel hesitant or afraid of failure. You may try something that doesn't work the first time. That is not a wasted effort; that is part of the technocurious stance we advocate. Each attempt teaches us something about the possibilities and limits of these tools.

But as teachers, we are not passive recipients of change. We are stewards of disciplinary knowledge, of classroom experience, and of the traditions of historical thinking. These commitments must shape how GenAI is introduced into our classrooms. We invite you to explore the terrain of GenAI with curiosity and care, experimenting and questioning, but with clear direction. We may not have chosen the fact that GenAI is present in education, but we can choose how we respond.

Through the Looking Glass: Cautious or Curious?

- *When it comes to GenAI in your classroom, where do you find yourself today?*
- *What makes you curious about experimenting?*
- *What are you cautious about trying?*
- *How might you hold both at once, using curiosity to guide exploration while letting caution shape the boundaries of your practice?*

Image Long Description

Figure 1.2: Photograph of a printed letter addressed to parents and guardians. The letter is on a plain white background with black text. The header reads "Dear Parents and Guardians," followed by a message from a teacher outlining a policy regarding the use of artificial intelligence tools in English classes. The letter explains the prohibition of AI tools such as Grammarly and ChatGPT in student assignments to ensure academic integrity and the development of authentic writing skills. It lists specific reasons for the policy, including building writing skills, maintaining integrity, and providing accurate teacher assessments. The document ends with a request for parents to sign and return an acknowledgment, and an open invitation for questions.

2. THE TECHNOCURIOUS STANCE: A FRAMEWORK FOR TEACHING WITH GENAI

In the fifth chapter of *Alice's Adventures in Wonderland*, the Caterpillar and Alice look at each other in silence before he finally asks, "Who are you?" Alice falters. She admits she is not sure anymore, that she seems to have "changed several times since this morning." The Caterpillar insists that she explain herself. Alice tries, but she cannot pin herself down. What follows is a recursive exchange of question, response, contradiction, and repetition that circles back to the beginning. The Caterpillar is not satisfied, and Alice cannot quite resolve who she is. This puzzling dialogue captures something important about teaching in times of change. Alice is not sure, but she is curious enough to stay with the conversation, to keep circling back even when the answers are unsatisfying. That insistence on recursive questioning (i.e., returning again and again to the same problem with slightly different words, tones, or perspectives) is as much method as it is whimsy.

Figure 2.1: Illustration from Kingsley's *The Water-Babies*.

It is not an accident that Carroll gave this role to an insect. Early nineteenth-century children's literature in Britain often featured insects and animals as interlocutors:—figures of strangeness, natural science, and moral allegory simultaneously (Kohlt, 2016b). Self-improvement is intimately tied to the transience of life and shaped much of the Victorian imagination, spilling into both children's stories and popular natural history. Insects, with their astonishing capacity for transformation, became models for this moral pedagogy. Charles Kingsley's *The Water-Babies* (1895), for example, details the metamorphosis of a dragonfly from something quite ugly into a winged creature as a parable of moral progress; in this text, along with others from the time period, children who failed to improve often found themselves turned into insects instead. Carroll's caterpillar participates in this tradition, presenting children with a creature that is both familiar and alien, who embodies a form of recursive questioning that mirrors both the uncertainty of self and the moral weight of change.

"But hang on, we're history teachers!" you cry. "What is with all this *Alice in Wonderland* stuff?"

Well, give us a minute! This book weaves literary metaphors throughout the text alongside primary and secondary sources that link with (mostly) obscure historical events from around the time of Carroll's publication in Victorian England. We continue this theme throughout the book, partly just for fun but also to help improve your rankings at the Tuesday pub quiz. But also, this is how we are positioning technocuriosity and our use of Chatty Geeps in history education. Like Alice, teachers working with GenAI are being asked, again and again, "Who are you? What do you make of this? How will you respond?" The Caterpillar-like recursive questioning, with its overtones of transformation and uncertainty, offers a metaphor for how we might engage GenAI, not with final answers but with curiosity that returns, rephrases, and revisits. Within this space, the technocurious stance takes an amphibious perspective (Lury, 2012), circling back through iterative questioning, probing both the limits and the possibilities of new technologies without presuming a finished form.

Technocuriosity, as we describe in this book, is not about being good with or at tech. It is about being oriented toward inquiry. It is a way of approaching GenAI through experimentation and critique and with the willingness to say, in the spirit of Alice, "I can't explain myself yet because I'm still becoming."

Through the Looking Glass: Who Are You?

The Caterpillar keeps asking Alice the same question: "Who are you?" She struggles to answer, circling back with different attempts, never quite landing on a final definition. Now we ask you: "Who are you, as a history educator?"

- *When you encounter new technologies like GenAI, what questions do you find yourself returning to again and again?*
- *How does recursive questioning (revisiting the same question from new angles) shape your stance as a teacher?*
- *Where might curiosity serve you better than certainty?*

Alice as a Technocurious Educator: Positioning the Framework

Alice is not always certain of herself, but she rarely stops asking questions. Even when she finds herself in bewildering circumstances—shrinking and growing, being scolded by animals, or playing games with rules that change midway through—she continues to push forward. She experiments and reflects. Alice is open, questioning, and willing to explore, making her the perfect model of what it takes to inhabit a world whose logics are unsettled.

This posture offers a useful metaphor for how educators might approach GenAI. In our experience, most teachers find themselves somewhere along a spectrum. At one end are technoskeptics, deeply cautious about AI's risks and often inclined to keep it at arm's length (Pleasants et al., 2023; 2024). At the other are technoromantics, ready to embrace GenAI as a revolutionary tool, convinced it will transform teaching and learning if we let it (Fitzpatrick, 2024; Watson et al., 2024). Technocuriosity sits in the middle. It resists binaries and embraces our position *in medias res*, working from within uncertainty (Allen et al., 2026; Allen & Hicks, 2025; 2026). To be technocurious is to try, test, and reflect; to treat GenAI as a live, unfolding problem space; and, for our purposes, to keep the goals of historical inquiry firmly in view.

In Chapter 1, we briefly introduced some of the key tenets that shape this stance. Here, we provide more detail about the four concepts (figuration, configuration, liveness, and speculative futures) and pedagogical commitments that define technocuriosity.[1] Think of them as overlapping lenses rather than steps:

- **Figuration** describes the metaphors and figures we use to make sense of technology. Wonderland itself is a figuration in the form of a world where rules shift and identities transform. In the classroom, asking students to treat Chatty Geeps as a "first draft assistant" or "naïve historian" positions these GenAI applications as figures that shape how students engage. Think of GenAI not just as a tool but as a character in your classroom. It is sometimes a tutor, sometimes a mirror, and sometimes a distraction. It performs roles and carries symbolic weight, shaping how students (and teachers) imagine and construct knowledge.
- **Configuration** is the way the conditions are set for GenAI use. Just as Alice must adjust to each new space in Wonderland, teachers configure prompts, routines, and classroom norms that structure how students interact with GenAI. Configuration is about designing and learning to play the "rules of the game." Beyond this, GenAI is always entangled with the systems it enters, which affect it in the form of district policies, platform rules, and classroom norms. These infrastructures configure how the technology can be used, and in turn, it is reshaped by how teachers and students interact with it.
- **Liveness** is working in the middle of things (*in medias res*). Alice is dropped into scenes already in motion; history teachers are encountering GenAI in the midst of ongoing curricular, institutional, and political pressures. There is no pause button. Liveness means we learn as we go, remaining in motion rather than stopping to work from a finished script. Teaching with GenAI happens in the moment, whether it's part of class preparation or a tool used during class. A prompt that fails, a student question that derails, a response that surprises; these live interactions matter. Meaning doesn't exist apart from use; it emerges in the messy process of trying, failing, and trying again.

1. These concepts and pedagogical commitments were "poached" or "borrowed" then reimagined (de Certeau, 1984) from the field of sociology. In particular, we build on the work of Celia Lury as it relates to inventive methods (Lury & Wakefield, 2012), amphibious sociology (Lury, 2012), and figures (Lury et al., 2022).

- **Speculative futures/shadow lives** refer to imagining what could otherwise be. Wonderland is full of shadow possibilities in the form of creatures who might have been something else, rules that could shift again. Technocuriosity asks teachers to speculate: "How might GenAI expand the scope of who gets remembered in history and why? What risks might its 'shadow lives' pose for equity, truth, or disciplinary rigor?" Every GenAI response also carries absences:— what it leaves out, avoids, misrepresents, or distorts based on its training data. These "shadow lives" are just as important as the outputs themselves. They give us a way to talk with students about silences, bias, and what future the technology might be steering us toward.

Alongside these concepts, technocuriosity also involves three pedagogical commitments that align closely with what teachers already do when they approach the unfamiliar:

- **Inventive methods** call for teachers to engage as participants, not observers. Teachers can't stand back and wait for someone else to define GenAI's place in the classroom. Being technocurious means trying it out, testing prompts, noticing what happens, and reflecting from inside the process.
- **Speculative ethnography** refers to technocuriosity's position as a type of action research, with imagination and creativity turned to 11 rather than 10 (Reiner, 1984). It is not simply studying what happens but also undertaking small experiments and asking questions. Just as Alice pays attention to how Wonderland itself is shaping her, teachers can notice how the medium of GenAI reshapes what counts as knowledge in their own classrooms. The question moves from "What does it say?" to "How does this medium shape what it can and cannot say?"
- **Live methods** treat working with GenAI as something happening in real time. Teaching with GenAI is never finished or final. It is live; its outputs are ongoing and ever-changing. A technocurious stance treats that unpredictability not as failure but as an opportunity to learn to work in the middle of things, to adapt, and to improvise.

For teachers, the important point is this: technocuriosity is a stance, not a skill set. It does not require being "good at tech." It requires being oriented toward inquiry, holding curiosity and critique together, experimenting with caution, and refusing to let others dictate the terms of engagement.

Speculative Ethnography for Teachers

Speculative ethnography in the classroom means treating teaching as both observation and experiment. Alice doesn't just observe Wonderland; she speculates about its rules. As the Caterpillar leaves Alice, he says, "One side will make you grow taller, and the other side will make you grow shorter." In response, Alice wonders to herself, "One side of what?" When she realizes the Caterpillar is referring to the mushroom, she breaks off a piece from both sides and cautiously takes a small bite of each. You can respond to GenAI in the same way by engaging in your own wonderings and small experiments:

Step 1: Try (Small Experiment)

Pick one technocurious move to test. Example: Ask Chatty Geeps to draft two alternate options for a primary source analysis scaffold you use with students. Use one with students to test its potential, and keep the other in reserve.

Step 2: Imagine (What If?)

Pause and speculate about the option you chose. What if I shifted the prompt? What if I used this routine with multilingual learners? What if I prompted Chatty Geeps to make the bias more visible rather than less?

Step 3: Reflect (Observe & Record)

Note what you saw in testing the option. What were the student reactions, confusions, and insights? Treat these observations like field notes.

Step 4: Return (Recursive Loop)

Circle back with a small change next time. Rinse and repeat.

This framing positions you as researchers of your own practice. You are not just using GenAI but testing, imagining, and reflecting on it in ways that contribute to professional growth.

Anchoring Technocuriosity in Discipline-Specific Applications

Alice is more than merely confused when the Caterpillar repeatedly asks her, "Who are you?" She is negotiating change. Since arriving in Wonderland, she has grown and shrunk, she has remembered and misremembered information about her past life, and she has repeatedly found herself unable to explain who she is. Victorian writers would have recognized this as more than nonsense. At the same time Carroll was writing, scientific books about natural history were being marketed to children (e.g., *The Fairy-Land of Science* (https://digital.sciencehistory.org/works/o6hvpac/viewer/4191b1p) [Buckley, 1878]), while fiction writers were commonly using the scientific process of metamorphosis as a metaphor for moral growth. The Caterpillar, not yet a butterfly but with the potential to be, is the physical embodiment of the uncomfortable truth Alice is experiencing: identity is fluid and it is a process.

History teachers are also confronted with unsettling questions in this moment of change. Who are we, as educators, in an age of GenAI? What do our practices mean when tools can instantly produce fluent, plausible historical narratives? While technocuriosity does not provide answers to these questions, it does offer a lens through which to think about them.

Figure 2.2: *The Fairy-Land of Science*

To understand what technocuriosity looks like in history education, we situate this book in conversation with what we believe makes our discipline distinctive and powerful. At its core, history is an inferential and evidence-based discipline (Lee, 2005; van Hover et al., 2021[2]). History is not just a recounting of the past. The past refers to everything that happened, while history is what we can responsibly claim about it. Because we cannot return to the past, historians work with what remains: records and relics, texts and images, artifacts and landscapes, the layered debris of human life. These traces are sources, not evidence. Sources yield evidence only when they are used to support, test, or revise a claim in relation to a historical problem. Calling something "evidence" means its relevance has been argued, not merely assumed, and no matter how enchanting they appear, sources do not speak for themselves. This is why the same document may function differently across investigations, and why the question being asked shapes what a source can credibly say.

A deep understanding of history then cannot be achieved by simply memorizing lists of names, events, and dates. First-order concepts such as knowledge of people, places, and events in the past are important (especially at the Tuesday pub quiz). These concepts emerge from who, what, when, where, and how questions, and they form the basis for historical narratives. This chronicling of the past is what makes up our history textbooks and often structures our standards-based courses. But even here, teachers know the limits of coverage. Students can "do" Jamestown five times and still struggle to explain why it mattered, how people understood what they were doing, or what changed because of it. The problem is often not that students lack information but that they lack tools for making meaning with information.

2. You may have noticed our selective citation practices in this book. This is intentional. Rather than reproducing what Andrew Abbott (2006; 2008) describes as a "star system" of citation where certain scholars or frameworks exert a kind of gravitational pull and must be ritually invoked to signal legitimacy, we draw from a wider constellation of scholarship that informs how historical inquiry and thinking are taught and practiced. Abbott reminds us that such star systems can harden into generational paradigms, shaping what counts as legitimate knowledge while pushing other productive ideas to the margins. Our aim here is not to deny the influence of canonical figures but to resist ventriloquizing a small set of familiar names in ways that rule out disciplinary play, critique, and experimentation:— especially in a book that asks readers to think technocuriously about how knowledge is produced, circulated, and authorized in the first place. So if, as you read, you find yourself wondering why we didn't cite so-and-so or that one thingamabob ... now you know.

But history is more than the answers to these questions. Historical narratives are shaped through second-order concepts, the disciplinary ideas that organize interpretation and constrain speculation. These historical thinking concepts include inquiry, chronology, causation, significance, evidence, continuity and change, historical perspective, and the ethical dimension of the past (Historical Association 2019a, 2019b, 2019c, 2020a, 2020b, 2020c, 2020d, 2021a; Seixas & Morton, 2013; Ford, 2014). These concepts are about beginning to explain events, judge their importance, and understand them in context. Significance, for example, is based on a judgment made in relation to a question and a purpose, which is why different groups in different times can reasonably argue over what deserves attention and why. When students assume history is just a photocopy of the past, they are often pushed toward adopting a "jigsaw" approach where the task is simply to locate missing pieces in the copy. A better disciplinary image would be made up of Lego; the available traces of the past can be assembled into different, defensible accounts depending on the questions being pursued and the standards used to judge arguments (Ashby, 2017). In this sense, history is a public tradition of debate, where accounts are assessed by how well they use relevant evidence in relation to specific inquiries.

Finally, historians rely on procedural knowledge based on the strategies and methods that put second-order historical thinking concepts into practice (Ashby et al., 2005; VanSledright & Limón, 2006). This is the "doing" of history: asking questions, evaluating the reliability of a source, weighing different kinds of evidence, tracing causality across contexts, comparing multiple perspectives, or constructing arguments as ethically as possible. The actions are not add-ons; they are the discipline itself. They also require dispositions, not just techniques; historians will need a respect for evidence, a willingness to revise preferred stories when the record does not cooperate, and a commitment to treat people in the past as fully human.

What historians do when they ask questions, weigh evidence, and debate interpretations are the same kinds of moves we want students to practice and learn in schools. The difference is in scale. Students are not mini-historians, but they are learning ways of thinking that can take them beyond just passing a test and into the application of the same conceptual and procedural tools used in the discipline. In this sense, history education is less about memorizing what historians have already concluded and more about engaging students in work and ways of thinking that can help them make sense of the past in the present.

This understanding of history education[3] is one reason why we also like the term "historical literacy." Historical literacy is both cognitive and dispositional. It involves learning that history is dependent on interpreting sources that do not provide unproblematic pictures of the past and learning to treat history as a way of seeing the world that orients us in time and space. Historical literacy includes a commitment to making the strongest arguments possible in relation to questions and presuppositions, an acceptance that we may be obliged to tell

3. A personal aside from David: This description is very much me channeling my experiences with the Schools Council Project History (https://www.schoolshistoryproject.co.uk) at King James' School, under the (not always) patient guidance of Mr. Hargreaves and Ms. Walshaw.

different stories from those we would prefer to tell, and an ethical recognition that people in the past deserve the same respect we accord ourselves. In practice, this calls for empathy as a disciplined stance, a kind of "virtual experience" of trying to understand another's position while recognizing the distance and difference of that position and without claiming their voice. History education can function as a contact zone where students learn to examine "the other" in ways that resist speculation, stereotype, and easy moral certainty while remaining accountable to evidence and context (Walsh et al., 2017).

What Historians Do (Maybe)	What we Want Students To Do	Knowledge in Play/ Disciplinary Concepts	Technocuriosity With Chatty Geeps
Ask questions that guide/frame investigations of the past.	Frame their own historical questions (with scaffolds and support).	Second order (inquiry)	Practice recursive prompting; re-ask Chatty Geeps in different ways to improve the disciplinary nature of the student's emerging questions.
Gather and organize evidence from sources (traces of the past like records and relics).	Work with historical sources (primary and secondary) provided by teachers or found through their own guided research.	First order (substantive) + Procedural	Ask Chatty Geeps to suggest possible evidence, to categorize, or to create search terms and boolean phrases to improve searches of databases.
Evaluate the reliability and perspectives of sources based on the context of the time.	Practice sourcing with scaffolding.	Second order (evidence, historical perspective) + Procedural	Treat Chatty Geeps's outputs as texts to source. Who trained it? What voices are absent? Ask Chatty Geeps to evaluate a student's source analysis based on a guiding question and source analysis scaffold.
Corroborate across sources to build stronger claims.	Compare and contrast evidence gleaned from sources, noting agreements, silences, and contradictions.	Procedural	Contrast Chatty Geeps's answers with textbook passages or historical sources and note (mis)alignments.
Construct causal explanations (why events happened, how they connect).	Identify and weigh multiple causes (long, short, trigger) and consequences of an event or events.	Second order (causation, continuity, and change)	Prompt Chatty Geeps for causal chains by themes (social, political, economic, religious, military), and critique how well it explains change over time.

What Historians Do (Maybe)	What we Want Students To Do	Knowledge in Play/ Disciplinary Concepts	Technocuriosity With Chatty Geeps
Judge and ascribe significance (i.e., what matters for whom and why).	Evaluate which events or people should be most significant, justify why, and develop criteria for significance to support their inferences.	Second order (significance)	Ask Chatty Geeps to ascribe significance and rank events; interrogate its reasoning and reconstruct significance through classroom debate.
Build evidence-based claims/ interpretations and test out counterclaims.	Make evidence-based claims, and stress test via counterclaims.	Second order (evidence) + Procedural	Generate a GenAI counterclaim (*The Civil War was not about slavery.*), and dismantle it with evidence.
Reflect ethically on the uses of history in the present.	Consider the ethical dimensions of history, including how we remember and represent the past and how different generations represent the past.	Second order (ethical dimensions of the past)	Identify omissions and misrepresentations in Chatty Geeps's output (race, gender, class, disability, indigenous histories), then discuss why those gaps are ethically significant.

Figure 2.3: Relationships between what historians do, what we want students to do, and technocuriosity.

The question for us, then, is more than "How do we add GenAI to history teaching?" We want to consider how technocuriosity can incorporate GenAI tools in ways that extend disciplinary engagement, applying procedural strategies that deepen our students' understanding of history without displacing disciplinary learning. In the second half of this book, we dedicate full chapters to deep engagement with ideas that more fully respond to this question.

Ultimately, technocuriosity and our suggestions for incorporating Chatty Geeps do not replace disciplinary literacy. The nature of history as a discipline already embraces uncertainty and debate. Our use of GenAI applies and extends the application of these skills in a new landscape.

Profile of a Technocurious Teacher

Across the examples in this book, a profile emerges. The technocurious teacher is:

- *Reflective: Attentive to how GenAI outputs align (or clash!) with disciplinary practices like sourcing, corroboration, and significance.*
- *Playful: Willing to try odd prompts or surprising framings then pivot into serious historical reasoning.*
- *Critical: Interrogates not just the outputs but the silences, omissions, and infrastructures that shape them.*
- *Collaborative: Brings students into the process, treating GenAI as one text among many, never as an oracle.*

What does this look like in practice?

- *They try things out: experimenting with prompts and adjusting when the first attempt falls flat.*
- *They treat failure as data: recognizing that a misstep reveals how the tool is configured.*
- *They model sourcing and corroboration: using GenAI outputs as artifacts to question, not as final answers.*
- *They juxtapose tools: comparing their own prompts and preferred platforms with "wrapped" district platforms, asking which option better supports disciplinary thinking.*
- *They bring students into the loop: letting them critique, revise, and refine GenAI outputs as part of historical inquiry.*

Technocuriosity is curiosity paired with critique, a willingness to play, and testing boundaries while keeping historical thinking at the center.

Growing, Shrinking, Adjusting

After her encounter with the Caterpillar, Alice discovers that nibbling opposite sides of the mushroom enables her to change her size at will. Sometimes she grows so tall she can barely fit in a room; other times she shinks so much she cannot reach the key she needs. Only through trial and adjustment does she learn to manage these shifts.

Teaching with Chatty Geeps can feel much the same. At times, the technology seems overwhelming. It's too expansive, too disruptive, and too large to be contained by the ordinary boundaries of classroom practice. At other times, it feels trivial. It's too small to warrant attention compared to the weight of real disciplinary work. The technocurious stance invites us to treat these observations and interactions as the work itself, adjusting scale, experimenting with perspective, and staying curious in real time.

Technocuriosity is a practice enacted daily. It is a stance you take when you reprompt to look for changes, when you pause to question an omission, or when you invite students to critique an output alongside historical sources. It is curiosity coupled with critique, enacted through recursive moves of testing, reflecting, and revising. This chapter introduced the framework of technocuriosity and offered brief glimpses of what it looks like in practice for history educators. The next step is to pause on the crucial question "Can or should I (mis)trust this tool?" before experimenting with strategies that incorporate Chatty Geeps. In Chapter 3, we explore how history teachers can adapt familiar source-analysis habits to GenAI, developing checklists and classroom protocols to decide what is useful, what is suspect, and what is simply noise.

Like Alice adjusting her size, teachers do not need to find the perfect scale for working with GenAI. We need only a willingness to grow, shrink, and adapt as the landscape shifts, staying oriented toward inquiry, rooted in the discipline, and curious about what is yet to come.

Through the Looking Glass: Growing or Shrinking?

- *When it comes to working with GenAI, do you ever feel like it is "too big" (overwhelming, expansive, hard to contain) or "too small" (trivial, limited, not worth the effort)?*
- *In what situations does GenAI feel oversized or unmanageable in your teaching?*
- *When does it feel underwhelming or irrelevant?*
- *How might you adjust your scale like Alice, growing, shrinking, or shifting your perspective to find a stance that supports historical thinking in your classroom?*

Technocuriosity is a practice enacted daily. It is a stance you take when you reprompt to look for changes, when you pause to question an omission, or when you invite students to critique an output alongside external sources. It is curiosity coupled with caution, enacted through recurring moves of testing, reflecting, and revising. This chapter introduced the framework of technocuriosity and offered brief glimpses of what it looks like in practice for literacy educators. [illegible]

[illegible]

[illegible]

3. CAN I TRUST THIS TOOL? EVALUATING GENAI THROUGH A DISCIPLINARY LENS

THE BEGGAR'S PETITION.

Figure 3.1: A cartoon of the Tichborne Claimant, satirising his efforts to raise funds for his legal and personal expenses in 1872 having earlier been made bankrupt. The cartoon was published in *Judy; or the London Serio-Comic Journal* on April 24, 1872.

In the final chapter of *Alice's Adventures in Wonderland*, Carroll stages a trial that is both ridiculous and telling. The Knave of Hearts is accused of stealing the Queen's tarts, but the proceedings are anything but orderly. Testimony from the witnesses is nonsensical, and they repeatedly contradict themselves before the Queen of Hearts interrupts the proceedings with her infamous declaration: "Sentence first—verdict afterwards." In this world, evidence does not matter. Carroll's satire exposed what many readers would have recognized in Victorian England: legal systems are often more about power and performance than truth.

A few years after *Alice* was published, real-life England found itself following a trial that was no less spectacular, if not quite so fictional. In the Tichborne Claimant case (1871–72), Arthur Orton, a butcher from Australia, claimed he was the long-lost Sir Roger Tichborne, heir to a vast English estate (Roberts, 2022). Thousands of working-class supporters believed him, despite overwhelming evidence against his claim, including that Orton looked nothing like Roger, could not speak French (a language Sir Roger was fluent in), and was unfamiliar with basic details of the family estate. Public sympathy for a working-class "everyman" and resentment of the aristocracy turned the trial into a spectacle. Illustrated newspapers ran endless coverage, sketching his likeness and caricaturing the proceedings with political cartoons that added to the drama. Satirical journals like *Judy; or the London Serio-Comic Journal* (https://en.wikipedia.org/wiki/Judy_(satirical_magazine)) mocked the Tichborne trial relentlessly, turning the proceedings into a form of popular entertainment. In the cartoon shared here, the Claimant's body and posture are exaggerated to highlight the absurdity of his story, demonstrating how public opinion was being shaped less by evidence than by popular narrative. For nearly two years, Londoners packed the courtroom to witness this performance, in which class resentment and public sympathy mattered more than proof.

The connection between the Knave's trial and the Claimant's is thematic rather than historical. Both remind us that evidence can be overshadowed by narrative and that verdicts can be shaped as much by performance and prejudice as by proof. Carroll exaggerated this for comic effect, but as we see in the Claimant's trial, nonsense has the ability to become courtroom reality.

The same dynamic is at play today. GenAI produces polished text outputs quickly, fluently, and confidently, offering a kind of "verdict first." After the fact, teachers or students must discover whether the content is accurate, biased, fabricated, or misleading. Schools, too, have moved ahead with GenAI on a "sentence first—verdict afterwards" basis. Districts are purchasing licenses, mandating classroom use, or banning it outright without enough evidence to decide whether GenAI improves teaching or learning. Like Alice watching the trial unfold, history educators find themselves caught in the contradictions of fluent claims amid missing evidence.

For us, this chapter begins with a simple but urgent question: how do we help students distinguish between persuasive nonsense and disciplined truth when working with GenAI?

History teachers already know the answer in principle. We source, contextualize, corroborate, weigh evidence, and debate significance. In the age of GenAI, the need to model and teach discernment has never been greater. Just as Alice eventually asserts herself in the face of arbitrary authority, declaring to her would-be jailors, "You're nothing but a pack of cards!" teachers must also reclaim their authority in evaluating new technologies. The verdict cannot precede the evidence.

Everything Is Important: Write That Down!

> *"What do you know about this business?" the King said to Alice.*
>
> *"Nothing," said Alice.*
>
> *"Nothing whatever?" persisted the King.*
>
> *"Nothing whatever," said Alice.*
>
> *"That's very important," the King said, turning to the jury. They were just beginning to write this down on their slates, when the White Rabbit interrupted: "Unimportant, your Majesty means, of course," he said in a very respectful tone, but frowning and making faces at him as he spoke.*

> *"Unimportant, of course, I meant," the King hastily said, and went on to himself in an undertone,*
>
> *"important—unimportant—unimportant—important—" as if he were trying which word sounded best.*
>
> *Some of the jury wrote it down "important," and some "unimportant." Alice could see this, as she was near enough to look over their slates; "but it doesn't matter a bit," she thought to herself.*

Carroll's satire in this passage exposes a courtroom where evidence has been replaced by performance and where the act of writing something down makes it appear meaningful, regardless of truth.

This type of comprehension is not far from what we see in GenAI outputs today. Programs like Chatty Geeps produce fluent text that performs knowledge, but fluency is not always truth. Like the Tichborne Claimant, Chatty Geeps is often a slick-talking imposter doing its best to convince you that it is the real thing. And, like the Wonderland jurors, Chatty Geeps and other GenAI chatbots currently have no built-in way to decide what matters most. They confidently generate information at speed without distinguishing between the important and the unimportant.

That burden falls to us. Teachers and students must act as the filter, following the principles of historians to decide what deserves to be written down. Without emphasis on these disciplinary actions, every output risks being treated like the King's nonsense, considered "very important" simply because it exists.

Realities of GenAI Outputs

One of the first things to understand about GenAI is that it is not an author in the traditional sense. It does not "know" history, nor does it draw on a stable archive of curated sources. Instead, it produces responses probabilistically, stringing words together based on patterns. In other words, large language models like Chatty Geeps are not sentient, wise, or all-knowing. As we suggested in Chapter 1, they are statistical next-word generators, spectacular guessers who are trained on immense quantities of training data. Every output is a kind of guess:— fluent, confident, and often persuasive but not the result of reasoning or evidence.

As a result, GenAI's style can be misleading. Its sentences read smoothly, the tone is authoritative, and the structure resembles academic writing. The fluency of outputs can appear "super" human, but they are nothing more than improv machines, just improvising, mirroring, and building on language patterns they have been trained on. They often privilege staying in character and maintaining agreeableness and narrative momentum over factual accuracy or safety. They can appear empathetic, agreeable, or wise, but in reality, they are only mimicking our own language patterns back to us. Fluency, however sophisticated or sexy, is not truth. Beneath the polish, the accuracy of these models is inconsistent. Teachers who have experimented with Chatty Geeps know this already. A prompt for a source analysis scaffold can produce a usable routine one moment and fabricate citations the next.

The issue is compounded by the opacity of training data, some of which is stolen, and the companies are being sued (Carson, 2025). Much of what fuels GenAI comes from the open web, scraped without permission, poorly documented, and uneven in representation. Its outputs reflect not only bias but also absence. Voices and perspectives are marginalized or erased. When GenAI offers a summary of the Reconstruction era, for example, its confident narrative may echo mainstream textbook framings while silencing Black voices and the complexity of local experiences. These silences are as important as its errors.

In this sense, GenAI mirrors Carroll's parody of justice and the spectacle of the Tichborne trial. In Wonderland, jurors mark evidence as "important" and "unimportant" blindly, without distinguishing between the two. In England, thousands rallied behind the Claimant, persuaded by his performance rather than by the facts. Likewise, when students or teachers encounter polished GenAI output, the temptation is to accept it at face value. Confidence and performance may overshadow the need to critique the evidence. Instead, teachers must demonstrate to students that not everything chatbots generate is equally valuable. Just as we ask students to source documents, corroborate claims, and weigh significance, we can model these same habits when using Chatty Geeps. Treating outputs as texts to be evaluated rather than answers to be consumed shifts the frame to inquiry. When students are not equipped with the tools to discern what is true and what is not, along with what matters and what does not, GenAI can act as a platform where verdicts are declared before evidence is weighed and performance substitutes for truth.

The Smooth Talk Illusion

Don't be fooled by what looks like slick, sexy, and knowledgeable chat. Fast, smooth, and stylistically polished communication can seduce us into mistaking style for substance. Scientists call this processing fluency (Schwartz, 2018): when something feels easy to read, hear, or follow, our brains are more likely to judge it as true, coherent, or even beautiful. That means when something sounds articulate and confident, we're more likely to believe it's true, even if it isn't. Fast, smooth, confident outputs from people or GenAI require caution because we're exquisitely sensitive to the feeling of fluency but rarely notice its source. The result is the subtle illusion of ease masquerading as accuracy. The next chart breaks down why "smooth talk" feels so convincing and how both people and machines can exploit our tendency to trust fluency over evidence.

The Human Analogy	The GenAI Parallel
Smooth/fast talk effect: We trust people who sound confident, fluent, sophisticated, even when the content or reasoning is weak.	**Fast GenAI effect:** GenAI systems produce polished and instant prose that feels authoritative. "Wow! It gave me so much writing. And it also said my question or prompt was great and worthy of a response!"
Biases at Play: **Halo effect** → One positive trait (e.g., confidence or a dazzling smile) shapes overall judgment. **Peripheral route to persuasion** → Style outweighs substance. **Processing fluency** → Smooth talk feels true.	**Biases at Play:** **Automation bias** → The machine must know because it is trained. "I mean, the calculator gets it right..." **Expertise heuristic** → Formal tone and even sentences with great grammar and punctuation must mean it is accurate.

Figure 3.2: Susceptibility to fluency over evidence

MAD HATTER ALERT: The "Don't Be Blinded by Fluency and Confidence" Checklist

*Evidence lens: Ask "Does the output **provide or point to evidence** or **just sound convincing**?"*

*Slow down lens: Ask "Am I agreeing because it **sounds good and/or agrees with me** or because **I verified it**?"*

*Corroborate lens: Ask "**Can I confirm this claim with another trusted source**?"*

*Substance over style lens: Ask "Is the **reasoning and argument evidence-based** or is it **just polished language**?"*

Building Your Own Evaluation Tool

Thoughtful history educators and students embody the opposite of Wonderland. We slow down before making decisions. But slowing down is not always easy. Protocols can help remind us of the value of slowing down and considering claims. The checklist below is designed to encourage teachers and students to slow down and thoughtfully examine Chatty Geeps's outputs as a source, carefully unpacking its responses before deciding if they are nonsense, half-sense, or good sense, then ultimately choosing to bin the output, modify it, or keep it based on evidence, reliability, perspective, and our own discernment.

The Checklist at a Glance[1] (3 lenses × 3 questions)

1. Attributing Purpose: Match the Output to the Task

    - Does this output align with my inquiry and standards?
    - Does it support historical thinking (sourcing, corroborations, causation, significance)?
    - Is the investment of time and energy worth the outcome? Why use Chatty Geeps for something when you could use a textbook, historical source set, or online educational resource?

    Quick Check – Action Items

    - Circle Verbs: Does the output support the teaching of the discipline (e.g., source, corroborate, compare, explain change and continuity)? Or is it just giving a summary or providing information?
    - Standards Match: Underline standards or compelling questions provided. If you can't underline it, what does that tell you about the output?
    - Worth the Effort (WTE) Test: Give it two minutes. Is this worth the effort to integrate into your instruction for students? Is it good enough to support or facilitate their learning of history?

2. Assessing Reliability: Pause for Evidentiary Warrant

    - Can the output be corroborated with a second independent source?
    - Are citations/sources real, accurate, and verifiable?
    - Is the content current and consistent with disciplinary expertise? Check dates, names, historiography, contested inferences, and interpretations.

    Quick Check – Action Items

    - Tagging: Mark claims as fact, interpretations, speculation, bias, or errors.
    - Source Hop (Lateral Reading): Verify key facts from a non-GenAI source.
    - Currency and Context: Check dates, editions, or challenge the source to see if it admits error uncertainty in debated areas.

---

1. This "3 lenses × 3 questions" checklist and decision chart builds on Gallagher, Swalwell, and Bellows's (2019) Pinning with Pause checklist for critically vetting online curricular curriculum (Table 1, p. 219), adapted here for evaluating AI-generated historical outputs with an evidence- and equity-focused lens.

3. Judging Perspectives: Probe for Voice, Damage, and Omission

    - Whose voices are included, centered, or omitted (e.g., class, gender, race, religion, indigenous perspectives)?
    - Does it reinforce dominant traditional narratives or offer/invite counter narratives and multiple causal explanations?
    - Could the output harm, offend, or marginalize students through stereotypes, decontextualized images, or by presenting "both sides" of the atrocity?

    Quick Check – Action Items

    - Perspective Sweep: Highlight whose standpoints are centered, add or push for missing perspectives.
    - Language Audit: Circle evaluative statements or adjectives; replace loaded language with descriptive evidence-based terms and phrasing.
    - Harm/Damage Check: Remove prompts/activities that could trivialize, shame, or deemphasize trauma and oppression.

Decision	**Triggers**	**Teacher Moves**	**Student Moves**
NONSENSE: Toss it in the rubbish bin! (Red)	Misaligned with standards and inquiry; unverified or fabricated citations; major factual errors and harmful framing	Don't use. Replace with vetted materials. The only potential application is to use it as a teachable moment on why we corroborate and verify.	Conduct an autopsy of the output. Identify the points of failure from a disciplinary lens (lack of evidence, can't corroborate, reified/unitary perspective).
HALF-SENSE: Small modifications might make it useful... (Yellow)	Basics are okay but incomplete; perspectives are missing, and there are weak connections to disciplinary thinking (describe, list, identify)	Triage it. Patch with historical sources, add counter narratives, and look for ways/verbs to stress historical thinking.	Annotate and identify claims, then add citations. Write a counternarrative or identify and add missing perspectives.
GOOD SENSE: This output could become a great teaching and learning tool. (Green)	Aligns with standards and inquiry; offers multiple perspectives and a chance to support, facilitate, and mentor disciplinary thinking; admits uncertainty	Use as a starter package to scaffold historical thinking, possibly pairing with other sources and resources. Be creative!	Apply a historical thinking scaffold (e.g., SCIM-C source analysis [see Chapter 9]) to stress test and produce claims with citations, reflect on the limits of the output, and gauge its value for supporting their disciplinary thinking.

Figure 3.3: Decision chart: Nonsense, half-sense, good sense

Testing the Tool: Make It Live in Your Classroom

By the end of the trial, Alice becomes increasingly frustrated by the nonsensical nature of the proceedings. She realizes that what she knows, her ability to engage in discernment, is more valuable than the results of the Wonderland trial and pushes the pack of cards, her would-be jailors, away before waking up from a dream. This is how Carroll's first Alice book ends. Our story doesn't end once we realize that what Chatty Geeps produces can be both important and not important, nonsense and good sense. Understanding the role of human discernment in using GenAI tools is a critical step for teachers who want to use these tools thoughtfully in the history classroom.

Just as we try to teach our students to evaluate sources for accuracy, purpose, and perspectives, we need to teach our students to do the same with GenAI outputs. They need to be put on trial. The spectacle of GenAI, with its polished sentences, authoritative tone, and speed of creation, can easily overwhelm the senses and discourage judgment and evaluation, but without evidence to back it up, the output may well be little more than persuasive nonsense.

Using the "3 lenses × 3 questions" checklist and the decision chart in conjunction with the steps laid out below allows both teachers and students to evaluate the outputs of Chatty Geeps.

1. Lay an AI-generated output alongside the protocol.
2. Evaluate output based on the content, focusing on specific historical thinking skills (e.g., evidentiary claims, corroboration, perspectives).
3. Decide whether it is nonsense, half-sense, or good sense, and record the decision.

Below, we provide examples of what this tool looks like in practice.

Example 1: Causes of D-Day (Oversimplification, Lack of Sourcing)

The Output

> Prompt: *Explain the causes of the D-Day invasion in World War II.*
>
> AI-generated response (excerpt): *The causes of D-Day were simple: to liberate Europe from Nazi control and end the war more quickly. The Allies launched the operation to strike at Hitler's forces in France. The invasion was successful because of careful planning, deception operations, and overwhelming Allied strength.*

Analysis Through the Protocol

- Purpose:
  - *Standards match?* → Partially. Yes, D-Day is required content in most state frameworks, but this reads like a Wikipedia-lite overview.
  - *Supports higher-order thinking?* → No. It lists causes as "simple," ignoring multiple layers (strategic, political, logistical, Soviet pressure).
  - *Verdict* → HALF-SENSE or NONSENSE, depending on class level
- Reliability:
  - No citations or evidence
  - Sweeping claim that "the causes were simple" signals oversimplification.
  - It could mislead students into thinking causation is linear and uncontested.
  - *Verdict* → HALF-SENSE, but only if paired with historical sources
- Perspective:
  - Missing Soviet perspective (their demand for a second front)
  - Missing French Resistance
  - "Overwhelming Allied strength" ignores German defenses and near-disastrous Allied losses on Omaha Beach.
  - *Verdict* → NONSENSE for upper-level work, HALF-SENSE for younger students with clear scaffolds

Example 2: GenAI "Creates" a Primary Source (Fabricated Evidence)

The Output

> Prompt: *Write a diary entry from a French civilian on June 6, 1944.*
>
> AI-generated response (excerpt): *June 6, 1944. Dear Diary, this morning I watched from my window in Caen as hundreds of Allied planes roared overhead. I heard that American soldiers landed on the beaches of Normandy and freed our town from German occupation. We cheered in the streets!*

Analysis Through the Protocol

- Purpose:
  - Compelling classroom activity; students love "voices of the past."
  - But this text presents itself as authentic when it is actually *fabricated*.[2] It presents fiction as fact.

2. See Chapter 11 for additional guidance on using first-person and perspective-based sources fabricated from Chatty Geeps.

    - *Verdict* → NONSENSE (for authentic source work). Could be HALF-SENSE only if reframed explicitly as creative writing or historical fiction, not evidence.
- Reliability:
    - Immediate red flag: Caen was not liberated on June 6; it remained under German control until July.
    - There is no citation; students could mistake this as a "real" diary.
    - The source fails basic corroboration as claims are fabricated.
    - *Verdict* → NONSENSE
- Perspective:
    - Erases the trauma of civilians (bombing casualties, destruction of Caen).
    - Romanticized "cheering in the streets" glosses over complexity.
    - Reinforces triumphalist Allied narrative without nuance.
    - *Verdict* → HALF-SENSE, usable if paired with authentic historical sources to show contrast; otherwise NONSENSE.

Example 3: GenAI Generates a DBQ Set (Omissions and Distortions)

Document-based questions (DBQs (https://www.dbqproject.com)) are designed to assess students' abilities to construct historical arguments by working with a curated set of authentic primary and secondary sources. Rather than recalling information, students must source, contextualize, corroborate, and use evidence from documents to support a claim. Because the integrity of a DBQ depends on the careful selection and authenticity of its documents, fabricated or poorly sourced materials fundamentally undermine the task.

The Output

> Prompt: *Create a DBQ (document-based question) set on the Marshall Plan.*
>
> AI-generated set:
>
>> Document A: Excerpt from Truman's 1947 speech (real, verifiable)
>>
>> Document B: Testimony from a Greek farmer praising American aid
>>
>> Document C: Statement from Soviet officials criticizing the plan

Analysis Through the Protocol

- Purpose:
    - DBQs align well with standards.

  - The skeletal structure is useful.
  - This could save teachers' prep time *if* the documents are real.
  - *Verdict* → HALF-SENSE (keep structure, replace bad docs)
- Reliability:
  - Truman's speech is real. Good.
  - The "Greek farmer" is fabricated (no citation, no archival reference).
  - The Soviet statement is vague; no sourcing is provided.
  - Half the set is unusable.
  - *Verdict* → The fake docs are NONSENSE; the set could be saved by adding real ones.
- Perspective:
  - AI determines what are considered pros and cons in the US–USSR conflict.
  - Missing European voices outside Cold War binaries (e.g., French skepticism, UK concerns about dependency).
  - Students are exposed only to cartoonish "praise vs. critique."
  - *Verdict* → It only makes HALF-SENSE; add missing perspectives.

Example 4: GenAI Summary of the Marshall Plan (A Usable Starting Point)

The Output

> Prompt: *Provide a summary of the Marshall Plan's goals and impacts, with citations to real sources.*
>
> AI-generated response (excerpt): *The Marshall Plan (officially the European Recovery Program) was announced by US Secretary of State George Marshall in 1947. Its primary goals were to rebuild war-torn Europe, prevent the spread of communism, and stabilize international markets. Between 1948 and 1952, the US provided over $13 billion in aid to sixteen Western European countries. According to historian Benn Steil (The Marshall Plan, 2018), the program not only revived European industry but also deepened Cold War divisions as the Soviet Union rejected aid.*

Analysis Through the Protocol

- Purpose:
  - *Standards match?* → Marshall Plan is a required Cold War topic (assuming this is true for your context).
  - *Supports higher-order thinking?* → Presents goals *and* consequences, setting up cause/effect analysis.
  - *Verdict*→ GOOD SENSE

- Reliability:
  - Dates (1947–52) and figures ($13 billion, sixteen countries) match standard references.
  - Citation check: Benn Steil's book is real and widely used in the field.
  - *Verdict* → GOOD SENSE (with spot-checking)
- Perspective:
  - Includes US aims and Soviet opposition
  - Acknowledges long-term division as well as recovery
  - Could still add European voices (British, French, German perspectives)
  - *Verdict* → HALF-SENSE (needs non-US perspectives but overall strong)

Ultimately, the goal isn't to make teachers suspicious of every GenAI output but to cultivate the same disciplined curiosity we bring to historical sources.

Through the Looking Glass: Potential Pitfalls

Chatty Geeps can be like the mirror in Through the Looking-Glass; *while it reflects the familiar shape of historical knowledge, the image it returns may be inverted, partial, or entirely invented. As history teachers, our task is to look carefully, noticing where reflection ends and distortion begins. As you experiment with AI-generated timelines, summaries, or sources, pause to consider the following common distortions in the mirror:*

- *Hallucinated Sources: Chatty Geeps invents fake citations, fabricated "diaries," or "documents" that never existed.*
- *Missing Perspectives: Dominant voices are centered, while marginalized voices are erased.*
- *Oversimplification: Causation is reduced to one-liners, and nuance disappears.*
- *Plausible Errors: Dates, events, or outcomes are subtly wrong but sound right.*

Have you observed these distortions in your own work with GenAI? How might you teach students to recognize when the glass is reflecting or distorting the past?

This Isn't Extra Work

Evaluating GenAI is not a new, additional task. It is simply a matter of applying the habits of historical thinking to a new kind of "source." Teachers already train students to source a document, corroborate across accounts, identify perspective, and distinguish fact from interpretation. As a result, implementing a GenAI protocol such as the one presented here uses the same disciplinary tools history educators have been teaching students to implement but applies them in a way that leads toward the development of critical AI Literacy by treating GenAI like any other source in the archive, subject to questioning, corroboration, and scrutiny.

CAVEAT! There will always be caveats. This protocol is not a recipe nor a set of rules that must be followed without question. Like any disciplinary scaffold, it is a thinking routine designed to slow us down, move through layers of inference, make judgment visible, and remind us to test claims against evidence. So feel free to adapt, borrow, or remix the questions for your own classroom contexts. In other words, add stuff that makes sense for you. For some tasks, you might only need the perspective lens; for others, you might need the reliability, or you might need to build a new lens! Use this tool flexibly to help develop habits of discernment and critique.

Flexible, Not Fixed

Just as Alice learns in Wonderland that rules can be arbitrary and absurd, we remind students that our evaluation framework should not be perceived as an absolute law. Rather, it is a guide to resist the Queen of Hearts' impulse to rush to judgment. It is a protocol designed to bend, adapt, and reshape.

- *This protocol won't cover every situation.*
- *Think of it as a map, not a mandate.*
- *Adjust the questions depending on your context.*
- *Feel free to create new disciplinary lenses.*

The key is building a habit of pausing and questioning, not following steps word for word. Be playful, be curious, and be speculative.

Building Habits of Critical AI Literacy

We define the idea of critical AI literacy as the capacity to (1) understand how AI works and where it came from; (2) appraise its strengths and weaknesses for specific scholarly and pedagogical tasks; and (3) evaluate the ethics of its use in different situations.

Once teachers and students have put a few outputs "on trial," the challenge shifts from winning a case to building lasting habits of discernment. Carroll's parody of justice in Wonderland works because it is so clearly absurd. But the danger is not confined to satire. In the Tichborne Claimant case, public sympathy and performance displaced evidence for years. The public spectacle held sway over proof. GenAI sits in that same gray zone. It produces polished, confident, and even seductive verdicts instantly while leaving the work of identifying evidence to us.

If this chapter's earlier sections offer a protocol for slowing down, this one is about turning that protocol into a habit. Teachers and students alike need routines that make discernment second nature. The goal is the cultivation of a stance based around a culture of pause, where evidence matters more than style.

For Teachers: Professional Habits in an Age of Fluency

History teachers already evaluate sources instinctively. We read a textbook excerpt and notice what's missing. We scan a historical source and weigh its reliability. The challenge with GenAI is to apply those same instincts to a different kind of document—one without a known author, archive, or provenance.

Consider the moment when Chatty Geeps drafts a bell ringer or exit ticket. The sentences look neat, and the grammar is flawless. But does it align with the inquiry of the day, or is it trivia dressed in polished language? In Carroll's courtroom, the King declared Alice's admission of knowing "nothing whatever" to be "very important," and the jury dutifully wrote it down. GenAI often tempts us into the same trap, leading us to write things down because they look official, not because they advance inquiry. We can reset the frame by taking a small pause, as little as two minutes, to circle the verbs and ask "Does this invite sourcing or just summarizing?"

Or imagine a handout generated in seconds, perhaps a timeline, a summary paragraph, a set of "model" responses. The fluency trap is especially dangerous here. What looks effortless can mask fabricated facts, missing citations, or errors that would embarrass us in front of students. Habitually cultivating does not mean being suspicious of every word; it more refers to a professional instinct to glance sideways, verifying dates, checking names, looking for receipts. If it passes, use it; if not, modify or discard.

These habits can be simple. You might consider keeping a "red flag log" in your binder so you have a running list of GenAI slips (fabricated citations, missing voices, biased framings) with notes on how you patched or replaced them. Over time, that log becomes both a vetted resource bank and a record of judgment. Or you might include a margin on all AI-generated handouts to act as an "evidence lane." This could be a space where sources and citations must be annotated. If it can't be annotated, it doesn't cross into instruction. These are not protocols so much as professional instincts externalized into routines that treat discernment as visible work.

For Students: From Shortcuts to Study Partners

If teachers are tempted to treat GenAI as a time-saver, students are even more likely to treat it as a shortcut. "Can you do this for me?" becomes the default question. The danger for students lies in Chatty Geeps's skill in creating fast, fluent answers that look like knowledge but mask the absence of reasoning, and the task for history educators is to redirect that impulse rather than simply prohibit it. Chatty Geeps can become a study partner or historical thinking mentor, but only if students develop the habit of questioning.

That habit can start small. Project a short GenAI paragraph on the board, and ask students to tag sentences as useful, suspect, or noise. In less than two minutes, they get practice sorting the wheat from the chaff. Over time, that triage can become routine. Or give students an AI-generated summary, and ask them to highlight one claim to corroborate with a textbook or historical source. It's not about producing perfect cross-references; what's more important is slowing down the rush of agreement: "This looks good, but let me check one thing."

You might consider implementing an "AI pause log" with students:— an alternate version of the red flag log. Each entry records the prompt, the output, and three quick moves: one fact corroborated, one missing voice identified, one claim discarded as noise. The log helps students visualize their discernment. Or you might choose to ask students to complete short reflection prompts as exit tickets. *What was useful in this output? What was suspect? What was noise?* Over time, the repetition strengthens instincts. Students begin to hear the inner voice of the historian: "Hold on... What's the evidence? Whose voice is missing?"

We are not creating students who are cynical about GenAI. Carroll's Alice does not reject Wonderland entirely. She plays along, she experiments, she listens. But at the trial, when performance tips into nonsense, she asserts herself: "Stuff and nonsense!" That is the moment of clarity we want for students: discernment about what deserves their trust.

From Critical AI Literacy to Disciplinary Literacy

Much of what is written about AI literacy emphasizes coding skills or digital safety. But for history educators, the challenge is different. We aren't turning students into programmers; we're turning them into critical readers of GenAI sources. That involves embedding GenAI into our existing routines and recognizing that Chatty Geeps is a text of uncertain provenance, not a neutral helper. Treating it as such by questioning, annotating, and contextualizing makes it less threatening and more useful.

This is why critical AI literacy is not an extra burden layered onto history instruction. It *is* history instruction. The Wonderland trial is absurd, but its lesson is serious. When verdicts come before evidence, truth is displaced by performance. Critical AI literacy means ensuring that evidence still rules the day in our classrooms.

Through the Looking Glass: Useful, Suspect, or Noise?

- *When you read an AI-generated response, how do you decide what to do with it?*
- *What makes part of an output useful for your teaching or for student inquiry?*
- *What signals that something is suspect and requires closer scrutiny?*
- *What do you recognize as "just noise"(material to discard)?*
- *How might your existing habits as a history teacher (sourcing, corroboration, weighing significance) guide these judgments?*

4. HOW TO USE THIS BOOK WITHOUT LOSING YOUR MIND

When Alice tumbles down the rabbit-hole, she notices cupboards, bookshelves, and even "maps and pictures hung upon pegs." She catches glimpses of knowledge as she falls, but she has no way to orient herself. The maps are there, but they do not tell her where she is going. Wonderland unfolds as both disorienting and inviting, a place full of possibility but without a clear path forward.

That image resonates with many history teachers, who feel similarly about GenAI. We glimpse resources, promises, and possibilities. Platforms offer premade tools. District initiatives hang policies in front of us. Colleagues share experiments or discuss fears. The landscape surrounding GenAI is crowded, but it is unclear in which direction we should travel. Like Alice, we fall past maps and charts without knowing how to read them or how they connect to the journey ahead.

How the Book Is Structured

This chapter is meant to be an orientation tool to help you better understand how to use this book, not a definitive instruction manual that addresses every possible question. Think of it more like an atlas that gives you directions for how to begin your own journey. Manuals assume there is a single, step-by-step route from beginning to end. Atlases, especially early modern ones, are different. They are designed for the user to start where it makes the most sense based on where they are located. And teachers, like explorers, know that maps are never neutral. They highlight some routes, hide others, and historically have included speculative guesses as much as settled knowledge.

The Waldseemüller map of 1507 (https://www.loc.gov/collections/discovery-and-exploration/articles-and-essays/recognizing-and-naming-america) is a perfect example. Housed in the Library of Congress, it is famous for being the first map to use the name "America," but it is also riddled with speculation. Continents are distorted, and coastlines are imagined. Yet, for its time, it was incredibly useful. It offered navigators and scholars a way of situating themselves, even if the details were incomplete or inaccurate. It served as both a guide and an invitation to discover more.

Figure 4.1: The Waldseemüller map

This book works the same way. The chapters that follow are not a precise manual that lays out a perfect itinerary for teaching with GenAI. Instead, they are sketches, outlines of terrain that will help you begin your own journey. Some parts may be immediately practical or even familiar to you. Others may feel less applicable to your own classroom context; some may even feel unfinished. That is intentional. Just as the Waldseemüller map combined observation with imagination, this book is designed based on both what we have learned through our experiences with Chatty Geeps and what we hope may be possible in live classroom contexts. Ultimately, we want to help orient you to teaching with a tool like Chatty Geeps without prescribing a singular route.

In this sense, the book is less like a fixed itinerary and more like the maps Alice glimpsed as she tumbled past: partial, suggestive, and waiting to be interpreted. You can choose to begin with the framing chapters, skip ahead to a classroom routine, or linger in the appendices. Similar to speculative maps with their blank spaces (and sea monsters), the book warns of risks but also leaves space for curiosity and invention. You don't need the entire map before you begin. Just pick a direction and start to explore.

Choosing Your Path

Maps are useful, but they don't walk the journey for you. Even with an atlas in hand, travelers still have to decide which way to sail, which route to trust, and how much risk to take. The same is true for history teachers working with GenAI. You don't have to have the perfect plan. You just have to know where you want to begin.

There is no single "correct" entry point. Sometimes the way forward will be obvious, and sometimes it won't be. That's not failure; it is the nature of exploration. For some teachers, it's as simple as trying a single prompt, just to see what happens. This simple start is known as "low lift." For example, you might ask Chatty Geeps to draft three possible exit ticket questions about today's lesson or to generate two counterclaims your students might encounter in a debate. Small prompts like these are a way of dipping your toe into the pool, testing how the water feels without diving all the way in.

For others, the next step might be generating a small classroom routine, a short activity designed to scaffold inquiry, help students practice sourcing, or spark discussion. A mid-level option here might be using GenAI to model sourcing questions or to generate multiple framings of a causal question for students to analyze. These routines don't require you to overhaul your teaching; they build on habits you already have, offering a chance to practice technocuriosity in motion.

For those ready for something more ambitious, you might experiment with building or adapting a custom GPT, which we detail in Chapter 8. This is an ambitious lift, but it is also the most flexible, as it guides you in designing a model tuned to your own classroom, inquiry arc, or disciplinary routines. Whether you share it with students or use it as a behind-the-scenes planning tool, this is where experimentation gives way to invention.

Wherever you begin, the key is movement. Curiosity and experimentation matter more than perfect execution. One step forward, even a small one, is better than waiting for the whole map to come into view.

Choose Your Path

Not sure where to start? Think of it like trying to follow signposts in Wonderland. There's no single correct direction, just different ways to begin.

Low Lift: A Single Prompt

- *Try one thing.*
- *A quick, no-prep way to see how Chatty Geeps handles historical text.*
    - *Example: Ask Chatty Geeps to summarize this historical source in three sentences a middle schooler could understand.*

Mid-Level Lift: A Classroom Routine

- *Implement a small classroom routine.*
- *Adds structure, builds on sourcing routines, and integrates Chatty Geeps into classroom practice.*

- *Example: Give students a short historical source, and ask Chatty Geeps to generate three possible sourcing questions. Students then choose which question best deepens inquiry and explain why.*

Ambitious Lift: A Custom GPT

- *Adapt or build your own GPT.*
- *Higher investment, higher flexibility, rooted in disciplinary concepts*
    - *Example: Create a "Historical Significance Coach" that walks students through ranking events from most to least significant and prompts them to explain their reasoning.*

In Part Two of this book, watch for suggestions labeled Low, Mid-Level, and Ambitious Lifts along the way. These are teaching ideas for how you and your students might engage with Chatty Geeps.

Tools to Use Along the Way

Alice's progress often depends on small objects that appear at just the right moment, whether it's a bottle labeled *Drink Me*, a cake marked *Eat Me*, or a mushroom that can make her grow or shrink depending on which side she nibbles. Each object is portable, mysterious, and slightly unpredictable, yet each becomes a tool for navigating uncertainty.

In addition to the Low, Mid-Level, and Ambitious Lift suggestions, we have scattered similar tools throughout this book. They aren't potions or mushrooms, but they are designed to be just as handy. Each is a small, adaptable routine you can pick up, carry into your classroom, and remix to fit your needs. Think of them as magical objects you can return to whenever you need a boost of curiosity or a quick structure to test an idea. These tools fall into three categories:

Through the Looking Glass Moments

These are reflection prompts built for you (although you could implement something similar with your students). They invite a pause in the action to notice what feels disorienting, what feels promising, and what might be missing. Think of them like Alice peering into the looking glass; they're not about getting answers but about taking a moment to step back and see things from another angle.

Mad Hatter Alerts

These serve as warnings and reminders about the limits of Chatty Geeps. They highlight when a model may hallucinate sources, oversimplify narratives, or reinforce bias. They're the red flags scattered across the path, helping you notice when nonsense is being passed off as sense. Like Alice at the tea party, you'll learn that some claims only sound logical until you stop and ask, "Wait, does that actually make sense?"

Curiosity Cards

Included in the appendices, these are short, ready-to-use classroom activities that spark inquiry or help students practice core disciplinary habits. Like Alice's cake, they come in small portions but can have outsized effects. You might pull one out to jumpstart a discussion, model recursive questioning, or get students thinking about sourcing in a new way.

Figure 4.2: Alice, the March Hare, the Mad Hatter, and the Dormouse at the tea party

None of these tools are meant to be prescriptive. You don't have to try them all or use them in a set order. You can adapt these tools for your own classroom context. If a Curiosity Card feels too small, expand it. If a Mad Hatter Alert feels irrelevant, skip it. Like Alice experimenting with mushrooms and potions, the point is not perfection but play. See what happens when you try, adjust, and keep moving forward.

A Most Curious Tea Party

When we think of the "madness" in Alice's Adventures in Wonderland, *the Mad Hatter's tea party comes to mind. But Carroll wasn't writing absent of cultural norms and understandings. In Victorian England at the time of the book's publication, "madness" was both a medical category and a culturally significant phenomenon (Kohlt, 2016a).*

The infamous Bethlem Asylum ("Bedlam") had, for centuries, been open to paying visitors who came to gawk at inmates. By the nineteenth century, "pauper lunatic asylums" were filled with overworked laborers. These laborers included hatters whose mercury exposure gave rise to the phrase "mad as a hatter." Carroll himself had family connections to asylum reform through his uncle, who served on the Lunacy Commission.

Progressive psychiatrists of the era promoted a "non-restraint" system: gardening, sewing, and even tea parties meant to mirror respectable society. Visitors could witness inmates performing civility at these carefully staged events. Reports about these asylums describe tea parties that look strikingly similar to Tenniel's illustrations of Carroll's Hatter and Hare. Even details in Carroll's characters reflect these associations. To have "straw in one's head" was a common signifier of madness, one Shakespeare used in King Lear. Tenniel's March Hare is drawn with literal straw on his head, visually linking him to Victorian asylum inmates. Carroll's readers would have caught the reference immediately.

For Carroll, the Mad Hatter's tea party was both parody and critique. For our purposes here, we understand the Mad Hatter as a reminder that nonsense can sound convincing, performance can stand in for truth, and sometimes everyone at the table plays along.

Curiosity as Compass

Alice rarely knows exactly where she's going. Despite her uncertainty, she is carried forward by her willingness to move, observe, and adjust. History teachers navigating GenAI don't need a master plan either. This book is not a manual with one right answer, but an atlas of paths. It is a collection of partial sketches and speculative routes. You do not need to see the whole terrain in advance to begin to roam. Try something small. See what happens. Reflect. Adjust. Curiosity, more than certainty, is the compass here.

5. PROMPT ENGINEERING AS PEDAGOGICAL DESIGN

Twas brillig, and the slithy toves
Did gyre and gimble in the wabe;
All mimsy were the borogoves,
And the mome raths outgrabe.

"Beware the Jabberwock, my son!
The jaws that bite, the claws that catch!
Beware the Jubjub bird, and shun
The frumious Bandersnatch!"

He took his vorpal sword in hand;
Long time the manxome foe he sought—
So rested he by the Tumtum tree
And stood awhile in thought.

Figure 5.1: The Jabberwock

Lewis Carroll begins *Through the Looking-Glass* with "Jabberwocky," a strange verse widely recognized as the most famous nonsense poem in English literature. It is filled with absurd words that sound meaningful because they follow the rhythms of the English language. The reader feels confident they understand, reading and nodding along until they realize that the whole thing is utter nonsense. The grammar and syntax are technically correct, but there is no meaning. The "Jabberwocky" poem provides a convincing display of how correct form can mask emptiness.

GenAI works in the same way. It arranges words, pixels, or patterns with grammatical precision without understanding what any of it means. When we type a prompt, Chatty Geeps will smoothly and confidently (but often superficially) reply with perfect structure and grammatically correct sentences. But like Carroll's "slithy toves," it may still be nonsense until it comes into contact with human discernment. A teacher, a student, or a historian must add context, evidence, or intent for this output to be viable.

Prompt engineering isn't a technical trick. It's a thinking routine and, for history educators, a pedagogical move. Each word of a prompt encodes purpose, perspective, and boundaries, guiding the model toward clarity instead of confusion. This chapter explores how to move from the creation of nonsense like "Jabberwocky" toward making sense through carefully crafted questions that generate meaningful, disciplinary outcomes. If Carroll's verse makes us supply the sense ourselves, prompt engineering is how we find the sense in Chatty Geeps.

From Nonsense to Sense

"Jabberwocky" leaves readers in a strange interpretive position, as Carroll gives us his verse in a traditional poetic form but without any real substance. And yet, most readers, especially students, can't help but start to assign meaning. "'Twas brillig" must mean something like "bright" or "brilliant." "Gyre" and "gimble" sound action-oriented; "mome raths" must be creatures of some kind. Even the rhythm of the lines pushes us to fill the void. We picture the Jabberwock, the "jaws that bite, the claws that catch," and the hero's gleaming "vorpal sword." Based on what we know of the rules of English, we start to build this creature in our minds despite never having seen one.

That instinct to supply meaning is precisely what we want from students when they engage with historical evidence, inferring, testing, and making meaning from fragments. Carroll's nonsense succeeds because it invites readers to do what teachers hope students will do with sources: notice patterns, question assumptions, and construct plausible explanations from incomplete information. The instinct to supply meaning from fragments is also deeply human and deeply historical. Medieval readers did it, too. Centuries before Carroll, artists and theologians filled the margins of illuminated manuscripts with beasts as strange as any Jabberwock: lions with halos, weeping pelicans, fire-born phoenixes, serpents with wings. These texts, called "bestiaries," are lavishly illustrated catalogues that combine natural observation, folklore, and moral allegory, in which authors both describe animals and ask readers to interpret them.

Among the most vivid of these is the *Aberdeen Bestiary* (~1200 AD), which features the illumination[1] of the dragon and elephant coiled across a field of gold shown in Figure 5.2. In the image, a sinuous blue dragon winds around a red-and-white elephant pressing its trunk toward the ground. The text that accompanies the image begins with a confident description: "The dragon is bigger than all other snakes or all other living things on earth" (Folio 65v).

1. Fun fact for the Tuesday pub quiz: The word "illumination" here is a noun that refers to the type of illustration. Illuminations were not merely decorative illustrations but material acts of craftsmanship. Artists applied gold leaf, silver leaf, or sometimes even powdered metals to the surface of the page so that it would catch and reflect light, transforming manuscripts into radiant objects where text and image seemed to glow from within (Beal, 2011).

Figure 5.2: Image of a dragon from the *Aberdeen Bestiary*.

The entry also includes a geographic aside. Dragons, it says, dwell in India and Ethiopia, where the heat nourishes their strength. But then the author pivots from description to interpretation:

> For lurking on paths along which elephants are accustomed to pass, the dragon knots its tail around their legs and kills them by suffocation. The Devil is like the dragon; he is the most monstrous serpent of all. ... The dragon lurks around paths along which elephants pass, as the Devil entangles with the knots of sin the way of those bound for heaven and, like the dragon, kills them by suffocation (Folio 66r).

In a few short lines, the scribe performs a linguistic sleight of hand, transforming what is meant to act as a nonfiction text into a moral treatise. The dragon's behavior becomes a parable of temptation and vigilance. The illustration reinforces the lesson that even beauty can conceal danger, and at this intersection, the "Jabberwocky" and the *Bestiary* meet. Both use language to generate creatures that never existed and rely on a predictable structure to make them feel real. Carroll's nonsense sounds sensible because it follows grammatical order; the *Bestiary*'s dragons feel credible because each entry follows an interpretive template made up of three consistent moves:

1. Description—physical form, habitat, behavior
2. Allegory—symbolic or moral significance
3. Application—instruction for the reader's conduct

It is, in essence, a medieval prompt structure, a way of turning the unknown into ordered knowledge.

For medieval scribes, moralized templates were a technology of learning. For modern teachers, prompts can be the same. In using Chatty Geeps, we, too, rely on structure to turn potential nonsense into learning. A well-crafted GenAI prompt mirrors the logic of a bestiary entry, setting context (description), defining the task (allegory), and clarifying expectations (application). *Write a short explanation of Reconstruction* is a description prompt. It is the context, broad and unbounded. *Explain how Black legislators in Reconstruction governments expanded democratic participation, citing one historical source* adds in a definition of the task and clarifies the expectation, performing the work of allegory and application in this example.

But prompting GenAI chatbots also requires discipline. Chatty Geeps, like the reader of *Jabberwocky*, works through probability. It positions words in the correct order, but does not understand why they matter or what they mean. When we type *Explain the causes of the Civil War*, the model recognizes the statistical associations of "Civil War," "causes," and "slavery," but it doesn't weigh evidence or wrestle with contingency. The work of the history educator is to provide that missing sense through well-organized prompt structures: defining purpose, setting context, specifying perspective, and identifying what counts as evidence. Moving from *Write about the Civil War* to *Write a paragraph explaining how economic and political tensions over slavery influenced the secession debates of 1860–61, citing at least two historical sources* requires the user to supply the meaning. In a way, this move makes visible the difference between "Jabberwocky" and disciplined prose. Each word in a prompt acts like a grammatical choice in Carroll's verse, shaping what follows and defining how readers (or in this case, machines) will make meaning. When we ask Chatty Geeps to "write a short explanation" versus "explain how," we are, in essence, authoring our own bestiary entries and defining the creature we hope to call forth. The next section turns from dragons to design, showing how the structure of our words shapes what the machine can (and cannot) imagine.

Dragons and the Power of Prompting

Amy's son Matthew discovered the power of prompting one evening at the kitchen table while playing with Chatty Geeps's image generator. His first prompt was a single word: Bob. *The exchange that follows demonstrates the robust capabilities of Chatty Geeps, the simplest form of prompting, and the absurdity of middle school boys.*

Bob

Hi! It looks like you just said "Bob." Could you tell me a bit more about what you'd like to create with "Bob"? For instance, do you have an idea of what kind of image you'd like, or maybe some details about Bob if he's a specific character or figure?

Dragon

Got it! Would you like a specific kind of dragon? For example, I could make it look fierce and powerful, gentle and wise, or maybe something a bit mystical with unique colors. Let me know any details you'd like to see—like its color, size, environment (mountains, cave, forest, etc.), or any special features (like multiple heads, wings, or horns).

Fat red fire

Made with the old version of image generation. New version available now.

Here's a chubby red dragon, complete with a fiery breath and a powerful, friendly look! If you want any adjustments or more details added, just let me know.

More fat

More fat

Made with the old version of image generation. New version available now.

Here's an even more rotund red dragon, with a massive, round belly and tiny wings for an exaggerated, almost cartoonish look. If you'd like further changes, just let me know!

Figure 5.3: Conversation about dragons between Matthew and Chatty Geeps

You can read the full chat here (https://chatgpt.com/share/68e65e09-72e0-8003-898d-7da7817f86f3). (It gets funnier!)

While this level of prompting created something Matthew was happy with, he was not really in control of the process in any way. Chatty Geeps maintained all the freedom. However, using a framework like the one presented here that includes a role, context, task, constraints, and output, we tried to harness the power of Chatty Geeps to create a dragon more similar to what Carroll and Tenniel had in mind when they pictured the Jabberwock.

Our prompt:

> You are a nineteenth-century fantasy illustrator commissioned to create a new dragon that visually echoes the Jabberwock from Lewis Carroll's *Through the Looking-Glass*, as illustrated by John Tenniel. Create an illustration of a dragon creature inspired by Tenniel's Jabberwock. You blend Victorian grotesque detail with mythic imagination and anatomical plausibility. The image will accompany a modern retelling of the poem "Jabberwocky," aiming to capture the eerie whimsy and surreal menace of Tenniel's style. The goal is to evoke familiarity while introducing original elements that feel freshly monstrous yet classically illustrated. Consider its body structure, texture, wings, scales, posture, and expression. Pay attention to linework, shading, and period-accurate illustration techniques. Constraints: Maintain black-and-white or sepia-toned ink illustration aesthetics, Use crosshatching and fine engraving-style texture typical of Tenniel's engravings, Avoid modern fantasy tropes (no neon colors, armor plating, or cinematic lighting). Include human-scale references (trees, figures, or ruins) for proportion. The creature must feel both absurd and terrifying.

Our result:

Figure 5.4: Chatty Geeps–created image of Tenniel's Jabberwock

We tried this prompt in multiple GPT's and got very similar results each time. Test it out on your own! It's not exactly right, but it is wildly different from Matthew's dragon.

Prompt Engineering as Pedagogical Craft

If the first half of this chapter explored *why* structure matters, this section turns to *how*. The following pages compile a range of frameworks, models, and examples with which history teachers can experiment when designing prompts for their own classrooms. The goal is not to master every tool or memorize a new vocabulary for prompt engineering. Instead, it is to see prompting as part of your process. We encourage you to consider prompts as tools that can be used to shape inquiry, test assumptions, and model historical reasoning through language.

When history teachers design prompts with intention and guided by context, constraint, and curiosity, they make inquiry visible. Every prompt reveals something about what we value, whether that relates to what we consider evidence, which voices we prioritize, or how we expect students to reason about the past. The practical routines that follow, such as dissecting the anatomy of a prompt or using the Prompt Compass to adapt disciplinary-specific models, are offered not as prescriptions but as instruments for thought. As we often like to say: beg, poach, steal, or borrow. Try one, test another, abandon what doesn't fit, and adapt what does. In this sense, prompting requires us, as history educators, to refine our own professional habits of mind. It invites us to articulate our reasoning and consider prompts as pedagogical tools.

Who Are You? Personalization and Prompting Behind the Scenes

Every conversation with Chatty Geeps begins with a prompt, and every prompt tells the model who you want it to be. In this short video, Amy pulls back the curtain on how "memory" works in Chatty Geeps and what thoughtful personalization can teach us about designing GenAI that remembers with care.

Access the "Setting up Memory" YouTube video by scanning the QR code or by visiting https://www.youtube.com/watch?v=lt6f9AjOs1w.

The Anatomy of a Prompt—Disciplinary Dissection

Before experimenting with complex models or frameworks, it helps to slow down and look closely at what a single prompt actually does. Like a historian dissecting a historical source, we can examine its parts, ask how each contributes to meaning, and notice what happens when a single element is missing. The chart in Figure 5.5 breaks down the core components common to effective prompting for history educators: role, context, task, constraints, output, and reflection. Then, it links each of these components to a familiar move in classroom practice. Think of it not as a formula to memorize, but as a map of possibilities. This structure provides a way to see how the language we use directs the logic Chatty Geeps uses to generate valuable outcomes.

Prompt Element	Function	Example
Role/Voice	Defines who Chatty Geeps should act as	*You are a historian mentoring a middle school student exploring Reconstruction.*
Context	Grounds the exchange in time, place, and purpose. Without providing specific context, responses tend to become overly general.	*The student is analyzing two newspaper editorials from 1865 about freedmen's rights.*
Task/ Inquiry	Frames the cognitive action	*Guide them to identify claims, evidence, and missing perspectives.*
Constraints/ Guardrails	Reduces hallucination risk and focuses the scope	*Only draw from the provided sources and clearly note when you are inferring.*
Output Format	Shapes how knowledge appears: dialogue, chart, short essay, historical think-aloud, scaffolded steps	*Respond as a dialogue coach with questions and hints, not full answers.*
Reflection Step	Encourages metacognition, asking questions, or addressing uncertainty	*End by asking: What new question could the student explore next?*

Figure 5.5: Anatomy of a prompt

Context Is Everything

Providing background and context (situating the prompt) may help to improve quality, reduce hallucination, and keep outputs historically faithful.

Context works at three levels:

1. *Historical setting*
    a. *Ground Chatty Geeps in a specific time and place.*
    b. *The narrower the frame, the truer the inference.*
    c. *Example:* You are analyzing American newspaper advertisements from 1943 about women factory workers.

2. *Classroom setting*
   a. *Let Chatty Geeps "know" who it's helping and why.*
   b. *This signals tone, complexity, and scaffolding needs.*
   c. *Example:* You are supporting a group of tenth-grade students preparing for a DBQ on Reconstruction.
3. *Purpose setting*
   a. *Explain what role Chatty Geeps should play (i.e., analyst, coach, or critic).*
   b. *When teachers set this background, Chatty Geeps aligns more closely with pedagogical and disciplinary aims.*
   c. *Example:* Act as a writing coach helping students strengthen historical evidence, not as a grader.

Prompting as Historical Thinking

Once we understand the inner workings of a prompt, the next step could be to connect those parts to the work of the discipline itself. The pedagogical moves at play when history teachers phrase questions direct students' thinking, what evidence they look for, and what they value as historical truths. In this sense, the idea of prompting is a modern extension of the scaffolds we already use to help students source, contextualize, and corroborate evidence. The table below links distinct aspects of historical thinking with specific prompt moves, showing how the language we use to engineer Chatty Geeps mirrors the language we already rely on to engineer inquiry.

Historical Thinking	Prompt Moves	Example
Sourcing	Ask about the author, audience, purpose, and/or perspective.	*What details in this document help you infer the author's motive?*
Contextualization	Anchor the event in broader patterns.	*What was happening during the time period this source was created?*
Corroboration	Compare accounts.	*Create a table that lists agreements and contradictions between Sources A and B.*
Significance	Judge what matters and why.	*Which event had the greatest long-term impact, and why?*
Causation	Link short- and long-term causes.	*Generate two chains of cause and effect for this event: economic and political.*

Figure 5.6: Prompting as historical thinking

Prompt Compass Framework

Seeing prompts through the lens of historical thinking reminds us that language isn't neutral. It's a design tool that shapes how reasoning unfolds. History habits like sourcing, contextualizing, and corroborating all depend on how questions are asked and how evidence is framed. The same is true when working with Chatty Geeps. Before bringing a prompt into the classroom, it helps to pause and examine whether it actually invites the kind of vibrant and resonant inquiry we want students to practice.

The Prompt Compass is designed to help teachers tune a prompt before bringing it into the classroom. It provides a moment to look around and ask, "What am I really trying to do here, and is it worth it?" In the rush to generate, revise, and teach with Chatty Geeps, it's easy to move quickly from idea to output. The Prompt Compass slows that process down. This quick reflective tool helps teachers develop prompts grounded in disciplined, inquiry-rich design.

Using the Prompt Compass isn't about getting it "right" the first time; it's about cultivating a professional habit of reflection. Before introducing a new task, read your prompt aloud, scan each dimension, and ask which areas feel strong, which are missing, and which could be made more precise. Be deliberate in your prompting. To familiarize yourself with the framework, run one of your classroom tasks through the Prompt Compass. Notice what it emphasizes or assumes, then make revisions if necessary.

The Prompt Compass works as both a mirror and a map. As a mirror, it helps teachers recognize how their own pedagogical choices (e.g., the verbs they select, the context they include, the constraints they set) reflect their disciplinary values and theories of learning. As a map, it helps chart a route through the unpredictable terrain of GenAI interaction. It embodies the spirit of technocuriosity, which relies on the willingness to slow down, question, iterate, and adjust while working within live systems that are constantly changing.

The Prompt Compass is less of a checklist and more of a practice that builds curiosity and care into the act of working with GenAI. In this sense, each dimension of the Compass can be seen as a point to stop for reflection. Together, they help us navigate the messy middle ground.

Dimension	Curiosity Nudge	Strong Prompts Do This	Common Rabbit Holes	Revision Moves
Purpose	Why am I asking this?	The goal is clear and linked to a historical thinking skill (e.g., sourcing, evidence, significance)	Prompts request recall or generic summaries	Add verbs of thinking: *analyze, evaluate, interpret, compare*
Precision	Does Chatty Geeps know what world it's in?	Provides historical context, role, and boundaries	Responds with anachronisms or generalities	Specify historical context and/or anchor in time and place: *Act as a historian analyzing newspaper reports from 1919.*
Perspective	Whose voice or viewpoint matters here?	Invites multiple perspectives; notes uncertainty	Ignores bias or missing voices	Add clauses like *Identify two contrasting perspectives and their sources.*
Process	How should the reasoning unfold?	Breaks the task into steps; models historical method	Delivers flat answers without process	Add sequencing, and scaffold the reasoning: *First identify claims; next locate evidence; finally assess reliability.*
Pedagogy	What will the learner actually do with this?	Encourages student agency; the GenAI questions, models, or reflects	Dominates or provides ready-made answers	Recast Chatty Geeps as mentor. Make prompts that ask the student to explain their reasoning.
Guardrails	How do I keep nonsense and madness at bay?	Limits hallucination; asks for citations and uncertainty labels	Fabricates evidence or conflates sources	Include checks: *Only use uploaded texts, and note when you are uncertain.*
Output Clarity	What shape should the answer take?	Specifies desired form (e.g., table, dialogue, SCIM-C outline)	Response is unfocused or verbose	Add format cue: *Respond as a two-column comparison chart.*

Figure 5.7: The Prompt Compass

MAD HATTER ALERT: Guardrails Against Hallucination

When it comes to GenAI, a few "guardrails" can help teachers and students slow down, check evidence, and separate sense from madness.

1. ***Anchor in source data.*** *Always provide a real text, excerpt, or image when asking Chatty Geeps to analyze. Never ask for "examples" in the abstract. Hallucinations happen most often when the model has nothing real to hold on to. Better to have it misread an actual source than invent one from thin air.*
2. ***Signal uncertainty.*** *Ask Chatty Geeps to tag which claims are based on evidence and which are inferred. Have it label speculation or low-confidence statements directly in its output. These uncertainty tags create visible checkpoints for student verification.*
3. ***Interrogate confidence.*** *Prompt the model to rate its confidence in each claim and to explain why. Require citations for high-confidence statements, then check them. (Spoiler: sometimes the citations are fabricated. That's part of the lesson.)*
4. ***Prompt for transparency.*** *Ask explicitly:* Where does this idea come from? *If the model claims a source, verify it; if it doesn't, assume it's blending phrases and probabilities.*
5. ***Limit scope.*** *Narrow time period, place, and theme. The tighter the context, the less likely the model is to drift into vague generalities or textbook-style summaries.*
6. ***Compare versions.*** *Run the same prompt two or three times. Compare how the outputs differ in tone, emphasis, or evidence. Use that variance as a teachable moment about bias, revision, and interpretation.*
7. ***End with reflection.*** *Close every Chatty Geeps exchange by asking* What should I check next to verify this? *or* What remains uncertain or debatable? *Have students (or Chatty Geeps itself) summarize open questions before moving on.*

Guardrails keep curiosity safe enough to wander and strong enough to question.

Through the Looking Glass: What Prompts Reveal

- *What does this prompt reveal about the nature and purpose of history?*
- *What might my students learn about historical thinking and the doing of history simply by watching me refine the prompt?*
- *What might my students learn about historical thinking and the doing of history by watching me use the prompt?*

Prompt Models for History Education

Now that we've explored how to tune prompts with the Compass, the next step is to put them to work. The following Prompt Models for History Education illustrate how different prompt designs can support distinct classroom goals, whether that is mentoring students through writing, scaffolding inquiry, or surfacing multiple interpretations of the past.

Each model offers a starting place. Begin by selecting a model that aligns with your learning goals. Perhaps one of the ideas below fits an upcoming unit, a skill you want to emphasize, or a type of student thinking you want to encourage. Then run it through the Prompt Compass to strengthen its context, guardrails, and reflection moves. Finally, test the prompt in Chatty Geeps and determine if the response models the kind of historical reasoning you want students to practice. If not, refine it.

Ultimately, these models are designed to help you see prompting as a flexible pedagogical act, a way to make historical inquiry visible and extend your classroom dialogue through the use of GenAI.

Model	Description	Classroom Example
Socratic Method Goal: Encourages reflective dialogue, critical reasoning, and student agency; aligns with self-regulated learning	Uses iterative questioning to deepen reasoning and self-regulation; Chatty Geeps prompts students to clarify, justify, and reconsider their claims	*You are a Socratic coach helping students analyze the causes of the Civil Rights Movement. Ask three probing questions that push them to use evidence and address counterarguments.*
Inquiry Model Goal: Scaffolds inquiry design, connects questioning to evidence, and models the iterative nature of historical research	Structures prompts around the C3 Inquiry Arc: compelling → supporting → source → argument → reflection	*You are guiding a C3 inquiry into the question, "Was the American Revolution inevitable?" Help students generate three supporting questions and identify historical sources to investigate each.*

Model	Description	Classroom Example
Mentor Model Goal: Supports formative assessment and self-regulation; humanizes GenAI as a partner for improvement, not automation	Frames Chatty Geeps as a supportive co-teacher who gives formative feedback, scaffolds writing, or models reasoning aloud	*You are a writing mentor reviewing a student's DBQ draft. Provide comments that focus on clarity of argument and use of evidence, not grammar.*
Disciplinary Model Goal: Reinforces disciplinary thinking; GenAI acts as a demonstrator of process, not a deliverer of answers	Grounds Chatty Geeps in historical heuristics (sourcing, contextualization, corroboration, causation, significance)	*Model how a historian would corroborate these two accounts of the 1919 Race Riots. Explain which evidence strengthens or weakens each narrative.*
Historiographical Model Goal: Models disciplinary meta-awareness; students see history as debate, not consensus, and learn to treat historians as evidence	Uses Chatty Geeps to identify or compare multiple interpretations of the same event, revealing how historical knowledge evolves.	*You are a historian analyzing how interpretations of the French Revolution's causes have changed from the nineteenth to the twenty-first centuries. Summarize at least three distinct schools of thought, cite representative historians, and explain what evidence or ideology shaped each interpretation. End by suggesting how a new historian might reinterpret the event today.*
Structured Academic Controversy (SAC) Model Goal: Fosters empathy, civility, and evidence-based debate; supports civic discourse and the C3 Dimension 4 skill of informed action	Frames Chatty Geeps as a neutral facilitator helping students research, argue, and then synthesize opposing historical positions	*You are a discussion coach preparing students for a Structured Academic Controversy on the dropping of the atomic bomb in 1945. Create balanced briefings: one set of arguments supporting Truman's decision and one set opposing it, each with cited evidence from historical and secondary sources. Then prompt students to summarize areas of agreement, disagreement, and potential synthesis.*

Model	Description	Classroom Example
Counterfactual Inquiry Model Goal: Strengthens causal reasoning, teaches contingency, and reinforces the difference between plausible inference and unfounded imagination	Uses disciplined speculation to reveal causation and contingency within evidence-based limits	*You are a historian helping students explore contingency in 1776. Construct a plausible counterfactual: How might the colonies have achieved autonomy without war? Identify which political, economic, and cultural factors would need to differ, and explain what this reveals about why revolution occurred as it did.*
Decolonizing Counterfactual Model Goal: Reveals structural inequality, fosters historiographical empathy, and connects inquiry to justice-oriented pedagogy	Combines perspective reversal with evidence-based imagination to expose power and absence in dominant narratives	*You are facilitating a decolonizing counterfactual on the Berlin Conference (1884–85). Ask students: What would Africa's political boundaries look like today if the conference had included equal representation from African leaders? Ground reasoning in period sources, and explain what this reveals about colonial power.*
Comparative Interpretation Model Goal: Builds students' ability to evaluate evidence, weigh interpretations, and trace historiographical change	Prompts Chatty Geeps to juxtapose historians' arguments or interpretations side by side for analysis	*Present interpretations of Reconstruction from historians W.E.B. Du Bois and Eric Foner. Create a comparison chart identifying their main claims, evidence, and contexts. Conclude with a question: Which interpretation better fits current scholarship, and why?*

Figure 5.8: Prompt models for history education

Complex Prompts and Historical Decision-Making Simulations

Some prompt models move beyond simple question-and-answer exchanges and begin to structure entire historical simulations. Simulations can be engaging experiences in a history classroom. They can help students grapple with uncertainty, competing perspectives, systems, and decision-making in ways that lectures or textbooks alone often cannot. At the same time, simulations are difficult to design well. Poorly designed simulations can oversimplify historical experiences, reinforce stereotypes, trivialize trauma, or drift into curricular violence (Evers & Wright-Maley, 2025). For this reason, we approach simulations cautiously and intentionally.

We are especially wary of activities that ask students to "pretend to be" enslaved people, Holocaust victims, Indigenous boarding school students, or other historical actors experiencing trauma. Simulations should function as tools for inquiry rather than reenactments (Hicks et al., 2025). In our own work, we avoid asking students to "be" people from the past. (For more on that discussion, see Chapter 11.) We do, however, sometimes use simulations to help students think about historical decision-making, uncertainty, evidence, and competing pressures. GenAI makes these kinds of simulations possible through complex prompting. Rather than asking Chatty Geeps for a single response, teachers can design layered prompts that structure information, sequence events, limit knowledge, and create branching decision pathways. In this sense, the prompt itself becomes part instructional design document and part simulation engine.

One example we developed focused on the Cuban Missile Crisis. Students worked in two groups: one representing the United States and the other the Soviet Union. In these small groups, students only encountered the intelligence briefings, political pressures, military concerns, and uncertainties available to their side at that historical moment. The simulation was intentionally designed to place students in conditions of escalating brinksmanship, where decisions had to be made with incomplete information, political pressure, and fear of catastrophic consequences. In this example, the instructional arc follows an "engagement first" approach (Ogle & Hicks, 2024), where students enter the crisis before fully mastering all of the historical content. Rather than beginning with a long lecture, students first experience uncertainty, make decisions, and compare outcomes. Only after this do they turn to sources, textbooks, and historical accounts to verify or revise their understanding. The simulation follows a sequence where students experience uncertainty → make decisions → compare pathways → investigate history → verify and revise their understanding.

As students moved chronologically through the crisis, Chatty Geeps revealed only partial information, asked students to justify decisions, introduced escalating tensions, and prevented hindsight from collapsing the uncertainty that historical actors faced. The goal was not to recreate Kennedy or Khrushchev psychologically, nor to "feel what they felt," but to help students recognize how historical decisions emerge through incomplete information, competing advisers, ideological framing, and fear of consequences.

Meaningful learning does not emerge automatically from immersion itself. Simulations require careful pedagogical mediation, structured inquiry, reflection, and debriefing. Chatty Geeps can help teachers create dynamic, branching historical scenarios, but thoughtful instructional design remains essential. The full Cuban Missile Crisis prompts are included in Appendix B for teachers interested in experimenting further. What other simulation ideas does this prompt generate for you?

Prompts Across the C3 Framework's Inquiry Arc

After experimenting with individual models, the next challenge might be to think systemically. How can prompts sustain inquiry over time rather than existing as isolated exchanges? Prompts are more than just one-time instructions.

NCSS's College, Career, and Civic Life (C3) framework (https://www.socialstudies.org/standards/c3) organizes social studies instruction around an inquiry arc that moves from developing compelling and supporting questions (Dimension 1), to applying disciplinary concepts and tools (Dimension 2), to evaluating sources and using evidence to make arguments (Dimension 3), and ultimately to communicating conclusions and taking informed action (Dimension 4). When used intentionally, prompts can anchor an entire inquiry arc, framing compelling questions, generating supporting ones, scaffolding source analysis, guiding argumentation, and closing with reflection. A single prompt can spark curiosity, unearth prior knowledge, or direct attention to evidence, but when linked together across a unit, prompts can chart the full arc of inquiry, from compelling question to reflection.

Within this arc, prompts can do far more than generate quick answers. They can help sharpen a compelling question so it invites argument rather than a simple yes or no answer; they can generate or sort supporting questions and model historical heuristics like sourcing, contextualization, and corroboration. They can scaffold writing and feedback, expand the range of historical perspectives students encounter, and reduce hallucination risk by embedding context and constraints. They can cultivate teacher reflection, helping us pause and consider what our questions reveal about how we understand and teach the discipline itself.

For example, teachers might begin an inquiry by using GenAI to test the strength of a compelling question. A prompt like *Suggest three versions of this question that highlight debate and evidence: "Did the New Deal end the Great Depression?"* can help refine how students engage with causation and contingency. The teacher's role here is to treat the AI's versions as mirrors. Each variation reveals assumptions about historical reasoning and interpretive focus.

In the supporting question phase, prompts can help students organize their thinking by category or concept. Telling Chatty Geeps to *Group these supporting questions by theme (economic, political, and social) to scaffold an inquiry on industrialization* encourages students to see relationships between ideas and recognize how historians organize evidence to pursue complex claims.

Prompts also play a key role in source work. When teachers feed Chatty Geeps a short excerpt from a historical text with the instruction *Demonstrate how to begin sourcing this letter using SCIM-C protocol*, Chatty Geeps can act as a tutor, walking students through the process, or it can create a mentor text. In this context, it becomes a model of disciplinary language and reasoning that students can analyze, critique, and emulate.

During argumentation and feedback, prompts can model reasoning or structure peer review. A teacher might prompt *Help a student weigh two historians' interpretations of Reconstruction. What evidence supports each?* The resulting output helps to interrogate how Chatty Geeps frames disagreement and which voices it amplifies or ignores. In this way, AI-assisted prompts make historiography tangible for students while giving teachers new ways to talk about bias and evidence.

Finally, prompts can help close an inquiry through reflection and self-regulation, inviting students to examine their learning processes and the limits of their understanding. You shift the goals of the exchange from mastery to metacognition when you share a final product with Chatty Geeps for evaluation and ask "What is missing from this analysis?" or "What questions did this inquiry raise about how history is constructed?" (We talk about how to use Chatty Geeps for assessment at length in Chapter 10.)

Function	Description	Example Prompt Play
Strengthening Compelling Questions	Use Chatty Geeps to test or refine compelling questions for inquiry units, ensuring they balance curiosity and conceptual rigor.	*Suggest three ways to reframe "Was the American Revolution inevitable?" to highlight causation and contingency.*
Generating Supporting Questions	Chatty Geeps can help unpack what students need to know to answer a compelling question, making inquiry visible.	*What smaller questions could help students decide whether the New Deal transformed government power?*
Modeling Historical Thinking Moves	Prompts can instruct Chatty Geeps to model how historians reason using sourcing, contextualizing, and corroborating.	*Demonstrate how a historian would corroborate these two sources about child labor in 1900.*
Building Language Scaffolds (text leveler)	Especially for multilingual learners, prompts can simplify, rephrase, or gloss historical vocabulary without losing accuracy.	*Rewrite this source excerpt at a Grade 8 reading level, keeping key historical terms intact.*
Enhancing Feedback Loops	Prompts can be designed to give feedback on student arguments, focusing on evidence and reasoning.	*Evaluate this paragraph for historical reasoning. Does the evidence support the claim?*
Expanding Interpretive Possibilities	Prompts can help teachers surface multiple perspectives or historiographical debates.	*Summarize two competing interpretations of the Marshall Plan: economic and ideological.*
Supporting Metacognition	Prompts can ask students to reflect on their process (e.g., how they approached a question, what assumptions they made).	*What did you notice about how you chose evidence? What might a historian challenge in your reasoning?*

Figure 5.9: Prompts to support vibrant and resonant historical inquiry

When integrated into the inquiry arc, prompts reveal their true pedagogical power for cultivating disciplined curiosity. They help students see that history is not a fixed story but a process of questioning, evidence-weighing, and meaning-making; the prompts can also become a record of practice that shows how our own questions evolve over time. In the next section, we turn to the design routines that help teachers experiment with these prompts in practice, discussing ways to write, test, and revise prompts that model the very same inquiry habits we hope to inspire in our students.

Through the Looking Glass: Dalí's Alice and the Art of Prompting

"It's no use going back to yesterday, because I was a different person then." — Alice

In 1969, the surrealist painter Salvador Dalí was commissioned by Random House to illustrate Alice's Adventures in Wonderland *(Popova, 2016). Long fascinated by dreams, distortion, and the instability of perception, Dalí was a perfect choice to pair with Carroll's equally disorienting world. He created twelve heliogravures[2]/prints that reimagined each chapter. Across them, Alice appears as a flowing silhouette, always shifting, never quite the same person twice. That interpretation echoes Alice's insistence that she cannot go "back to yesterday." Each encounter remakes her as each brushstroke reimagines Carroll's text.*

Figure 5.10: Dalí 's Heliogravure in the chapter "The Lobster Quadrille"

For teachers working with GenAI, this reflection feels familiar. We, too, are living in a moment defined by transformation, inhabiting a new, unknown world from which we cannot return. Prompting, like painting or reading, is iterative. Each revision reframes what we see and who we are as thinkers. When we adjust a prompt's tone, add context, or reframe evidence, we aren't just refining instructions; we are revising ourselves. Inquiry is an act of becoming, like art.

---

2. Our copy editor wanted us to remove heliogravures but we decided to use this opportunity to insert a little more pub quiz knowledge. Heliogravure is an intaglio printmaking process made up of two processes, creating the surface and printing the image with a copper plate. To learn more, visit https://www.youtube.com/watch?v=NEfojvBwFTo.

Consider one of Dalí's Alice illustrations alongside a "prompt makeover" routine (explained below).

- *How do historical context and perspective shape what Dalí saw in Carroll's text?*
- *How does iteration, whether in art, inquiry, or prompting, change what we can understand?*
- *What does Alice's insistence that she "cannot go back to yesterday" mean in a world where GenAI tools reshape thinking itself?*

Dalí's Alice reminds us that both art and teaching are interpretive acts. We can't return to the classroom we knew before GenAI. Each revision, each prompt, each act of reflection shapes us. Our task is to continue to play and see what new forms of meaning might take shape.

Evaluating a Prompt: A Prompt Makeover Routine

As with all good inquiry, the work of prompting extends beyond a single question, evolving through testing, reflection, and revision. The most powerful prompts are rarely written once; they're reworked, tried, and tuned until they invite the kind of historical thinking we want students to practice. These routines encourage teachers to write and test prompts as experiments, using iteration as a form of professional learning. For example, the Prompt Makeover Routine is designed to help you see how small changes in language can shift the reasoning that follows and how, through play and revision, we can model the same curiosity and adaptability that define our discipline.

The process is simple:

1. Write a quick prompt for a history lesson (something you might actually use in class).
2. Check it against the Prompt Compass (Figure 5.7). Which dimensions are strong? Which are missing or underdeveloped?
3. Revise it once. Make a small, intentional change. For example, add a role, clarify context, tighten scope, or introduce a reflection step.
4. Reflect. What changed about the kind of thinking the prompt now invites?
5. Revise again. Does the new version get you closer to the kind of inquiry you want? Or does it reveal that the prompt itself needs to be abandoned and reimagined?

In *Alice's Adventures in Wonderland*, Alice insists, "It's no use going back to yesterday, because I was a different person then." Prompt design works the same way, with each iteration revealing a slightly different result. It is up to the user to make adjustments.

Consider the following prompts:

> Prompt A: *Explain how the Industrial Revolution changed society.*
>
> Prompt B: *You are a history mentor helping a student compare working conditions in 1840s Manchester with modern global factories and the gig economy. Use evidence from the provided sources to do a think-aloud that helps students identify changes, continuities, and causes of those patterns.*

Which prompt would you rather give to Chatty Geeps? Which would lead your students deeper into evidence and interpretation? Run both through the Prompt Compass, test them, and see what emerges. Then discuss. Which prompt better models the kind of historical reasoning you want your students to practice? How might Chatty Geeps's response itself become material for student critique or revision?

The point is not to find the "perfect" prompt but to watch how framing changes reasoning. Prompting is a tool for noticing how inquiry unfolds, but it cannot replace thinking. As you iterate, you begin to notice that effective prompts are continually helpful when they are constructed with the discipline as their through line, with historical methods and heuristics as the rails that keep curiosity on track.

When practiced this way, prompt engineering shifts toward a technocurious mindset, with each prompt becoming a small expression of a teacher's theory of learning, reflecting thoughts on what counts as evidence, how thinking should be scaffolded, whose voices matter, and what forms of questioning lead to understanding. When we build prompts through the lens of historical thinking, GenAI stops feeling like magic and starts functioning as a mirror reflecting back the values and methods of our craft. Though Chatty Geeps does not always provide accurate information, in the hands of a curious and critical history teacher, it can be a powerful tool.

Other Prompting Frameworks to Play with

The 3Cs Framework:

- *Context + Constraint + Curiosity*
- *To support a historical inquiry*
- *Example:* Within the context of post-WWII Europe, identify two contrasting explanations for the start of the Cold War. Focus only on 1945–49.

The Prompt Sandwich:

- *Set the scene. → Ask the question. → Define the role.*

- *Example:* You are mentoring a student studying letters from freedmen after 1865. Ask three questions that could guide their analysis.

The Think–Test–Tweak Cycle:

- *Treat each prompt like a mini design experiment.*
- *Step 1: Draft the prompt.*
- *Step 2: Test it. What kind of reasoning does GenAI produce?*
- *Step 3: Tweak it for alignment with historical methods.*

The Disciplinary Spine:

- *Map prompts directly onto second-order concepts: evidence, causation, change, continuity, significance, and perspective.*
- *Keeps the prompts grounded in the language of the discipline.*

Holding Our Practice to the Looking Glass

At the end of *Through the Looking-Glass*, Alice finds herself reading "Jabberwocky" from the wrong side of the mirror. The words have been printed in the book backward, and to make sense of them, she has to turn the book toward the mirror and read the reflection. At that point, the letters fall into place and she can read the text, but she still doesn't understand what it means. As the vignette ends, Alice says, "It seems very pretty ... but it's *rather* hard to understand. ... Somehow it seems to fill my head with ideas—only I don't exactly know what they are!" Carroll adds that Alice "didn't like to confess, even to herself, that she couldn't make it out at all."

Chatty Geeps, like Carroll's poem, speaks fluently, but while it has the appearance of sense, it is down to the reader to create the meaning. When mastered, prompts become mirrors that help us reveal interpretable responses. In this sense, history teachers are learning to read from the other side of the mirror. But like Alice, when we hold our existing practices up to the mirror and look more closely at them, it's rather hard to understand. Chatty Geeps may be able to fill our heads with "pretty ideas," but we may or may not know what they mean or what to do with them. Our work with prompts shows us more than what Chatty Geeps can do. Each time we write or revise a prompt, we get a clearer sense of what we value in history teaching and where we might need to grow. In a way, we're writing our own bestiaries. We are trying to describe and interpret the strange new creatures that live in this age of GenAI, giving shape and meaning to what might otherwise feel like nonsense. Chatty Geeps doesn't give us answers any more than "Jabberwocky" does. But when we use it thoughtfully, it can help us see what is happening through the work of teaching. The mirror doesn't do the thinking for us. It just shows us where to start.

Through the Looking Glass: Reflections on Practice

- *What does one of your recent prompts reveal about what you value in history teaching (e.g., about how you define evidence, voice, or significance)?*
- *Where, in your own classroom, do you see the "bestiary" impulse, the need to describe and interpret the strange new creatures of teaching with GenAI?*
- *When you hold your work up to the mirror, what do you notice about your own curiosity? What remains unclear, and where might you look next?*

Image Long Description

Figure 5.3: Dialogue between GenAI and student about creating an image of a dragon. Prompt 1: "Bob". Response: Hi! It looks like you just said "Bob." Could you tell me a bit more about what you'd like to create with "Bob"? For instance, do you have an idea of what kind of image you'd like, or maybe some details about Bob if he's a specific character or figure? Prompt 2: "Dragon". Response: Got it! Would you like a specific kind of dragon? For example, I could make it look fierce and powerful, gentle and wise, or maybe something a bit mystical with unique colors. Let me know any details you'd like to see—like its color, size, environment (mountains, cave, forest, etc.), or any special features (like multiple heads, wings, or horns). Prompt 3: "Fat red fire". Response is an image of a dragon that is small, chubby, and red. It is breathing fire, with a friendly appearance, featuring a scaly texture and curved horns. Prompt 4: "More fat". Response is an image of a dragon that is larger, rotund, with a more exaggerated appearance, tiny wings, and a comical, cartoonish look.

6. DESIGNING BACKWARD WITH CHATTY GEEPS: RETHINKING LESSON PLANNING WITH AI

> "It's very good jam," said the Queen.
>
> "Well, I don't want any *to-day*, at any rate."
>
> "You couldn't have it if you *did* want it," the Queen said. "The rule is, jam to-morrow and jam yesterday—but never jam to-day."
>
> "It *must* come sometimes to 'jam to-day,'" Alice objected.
>
> "No, it can't," said the Queen. "It's jam every *other* day: to-day isn't any *other* day, you know."
>
> "I don't understand you," said Alice. "It's dreadfully confusing!"
>
> "That's the effect of living backwards," the Queen said kindly: "it always makes one a little giddy at first—"
>
> "Living backwards!" Alice repeated in great astonishment. "I never heard of such a thing!"
>
> "—but there's one great advantage in it, that one's memory works both ways."
>
> "I'm sure *mine* only works one way," Alice remarked. "I can't remember things before they happen."
>
> "It's a poor sort of memory that only works backwards," the Queen remarked.

In this scene from *Through the Looking-Glass*, Alice meets another queen, the White Queen.[1] Throughout this chapter, the queen seemingly remembers things before they happen, prompting much confusion from Alice. The claim that she lives backward, remembering things before they happen, feels absurd, but it may have more in common with the way teachers plan than we would commonly think.

The Understanding by Design (UbD) curriculum framework asks educators to do something that sounds simple but is strangely disorienting: imagine evidence of learning that doesn't yet exist. Design backward from the desired outcome, then work forward to plan the experiences that will produce it. It's a disciplined form of what the White Queen describes, requiring us to see the future and reason our way back to the present. Designing instruction in this way is disconcerting, as if you are living in two temporal registers at once, inhabiting both the present reality of today's classroom and the speculative space of what students may be able to understand as you move into the future. Yet, as the queen suggests, there is an advantage to developing a memory that

1. A personal aside from Amy: Carol Channing's performance as the White Queen in the CBS 1985 made-for-TV version of *Alice's Adventures in Wonderland* (yes, the one with Ringo Starr) is, in my view, unmatched. You can read more about this and see what I mean here (https://decider.com/2016/12/09/today-in-tv-history-alice-in-wonderland). This is not an objective claim about adaptations. It is, however, correct.

works both ways. It's how we learn to see instruction as more than just a sequence of activities. In this sense, the queen's "nonsense" logic captures something essential about the craft of lesson planning. It is both imaginative and procedural, anchored in future thinking but enacted in the present.

This paradox can also make designing with Chatty Geeps feel uncanny. We are planning learning that hasn't happened with a partner that has only a limited memory (that you must provide), no sense of time, and no lived experience of learning. When co-creating lesson plans with Chatty Geeps, it becomes the teacher's task to provide the temporal discipline that GenAI lacks, such as the ability to anticipate and recall, to imagine forward and design backward. This is why, even in the age of being able to produce lesson plans immediately, a concept like backward design still matters. It provides a stable structure for working within an unstable technological landscape. Our collaboration with GenAI is grounded in something enduring through the identification of what students should understand and be able to do, the collection of evidence that will demonstrate that understanding, and the design of experiences that make learning possible. Without this scaffolding, GenAI becomes like the Queen's promise of "jam tomorrow," always just out of reach, endlessly promising but never quite there.

If Alice's conversation with the queen captures the confusion of living backward, it also points to its power. The queen's memory that works both ways is, in pedagogical terms, the essence of reflective design. Teachers must always look backward at what worked and what didn't while also projecting forward, imagining what might happen with this group of students next time or later in this unit. Backward design turns that reflection into structure. It gives us a way to turn the looping temporality of teaching into deliberate design.

Ada Lovelace and Ordering the Machine

The White Queen showed us what it looks like to live backward. Similarly, like backward design in teaching, mathematician Ada Lovelace's work began with the ends she believed a machine could achieve: to reveal the patterns of thought hidden within numbers. In practice, Lovelace also showed us what it looks like to design forward, retaining her imagination, embracing a systematic approach with a sense of wonder.

In 1843, Lovelace translated and annotated a technical paper on Charles Babbage's proposed analytical engine, a mechanical device that would have filled an entire room with cogs and wheels (Wolfram, 2015). While Babbage imagined a machine that could process numbers, Lovelace saw something far more radical. She recognized that the engine could manipulate not only numbers but symbols. Its operations could, in theory, be used to generate music, text, or images if only the rules were described clearly enough. "Note G" is a section of her 1843 paper that contains the first published computer program in the form of an algorithm for calculating Bernoulli numbers on Charles Babbage's hypothetical analytical engine. Lovelace grasped, almost a century before the first computer was built, that creativity and structure are not opposites. She understood that the art of design works in conjunction with the careful arrangement of rules that allow for surprise.

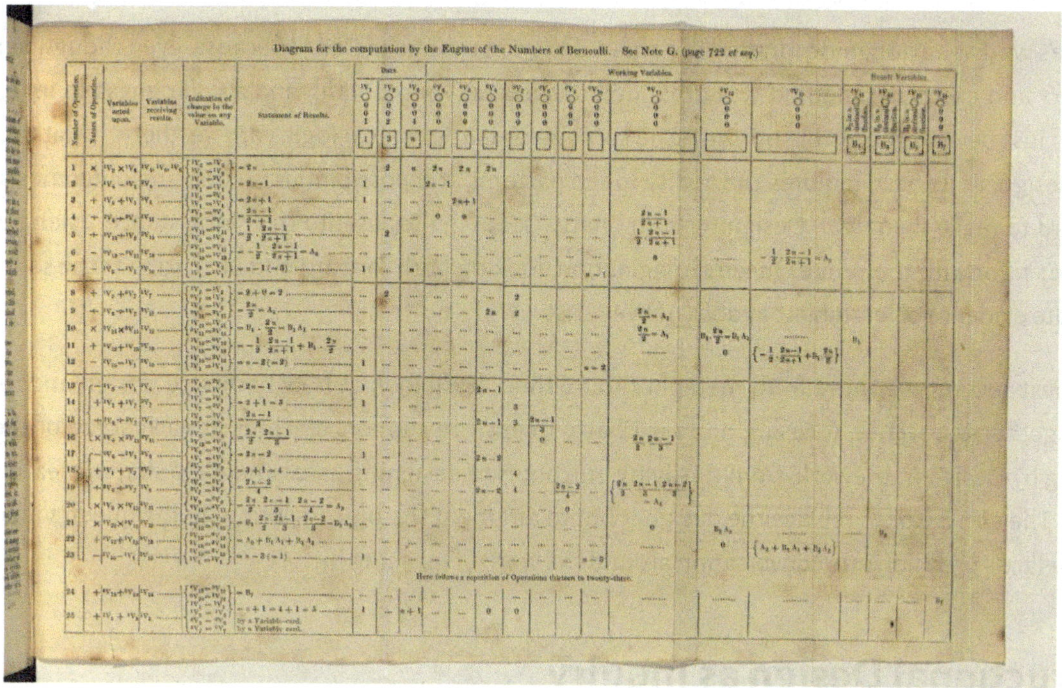

Figure 6.1: Note G by A. Lovelace

Instructional design, too, is a balance between what we can plan and what we must leave open. Like Lovelace, teachers are constantly writing algorithms from their experience sequencing classroom activities, developing formative assessment tasks, and making classroom management decisions. Every lesson plan is a small program that anticipates how learning might unfold and leaves room for what cannot be predicted. When we design instruction with Chatty Geeps, we find ourselves inhabiting Lovelace's world more directly than ever before. It is a machine that works only when we order it to. We must define the logic, the parameters, and the purpose of its operations. The quality of its output depends entirely on the precision of our prompts. But the most important work of design still happens before the prompt is written. It happens when we decide what matters, what counts as understanding, and how we want students to demonstrate it.

Backward design and Lovelace's algorithmic thinking share a fundamental orientation. Both begin by defining the desired output and then work backward to specify the inputs and processes that will produce it. In other words, both require us to imagine what does not yet exist. When teachers use Chatty Geeps to plan, they are not simply "asking" the tool for ideas; they are engaging in a form of algorithmic authorship, shaping the logic of the task and the reasoning of the model through their prompts. In other words, while Chatty Geeps executes commands, innovation is left up to the teacher. It can remix, reframe, and reassemble knowledge, but it cannot decide what is worth learning or why it matters. That remains the teacher's sole discretion.

Lovelace's work is an apt guide for thinking about design in the age of GenAI because, even though her process was strict and exact, her method was grounded in speculative reasoning and a willingness to imagine possibilities just beyond the visible horizon. She combined mathematical logic with aesthetic intuition. Instructional design, at its best, requires the ability to see patterns and possibilities, to imagine outcomes and work backward to make them real. Designing with Chatty Geeps invites us into this same stance, asking us to pair Lovelace's rigor with the White Queen's imagination, looking both forward and backward at the same time to anticipate evidence of learning that does not yet exist.

In the next section, we move from metaphor to method. We'll look at how backward design and GenAI can work together in practice, focusing on how Chatty Geeps can support teachers at each stage of the Understanding by Design framework, from clarifying outcomes to designing assessments to shaping learning experiences. The aim is not to replace the designer but to augment the process, bringing the queen's imagination and Lovelace's precision into conversation around the work of teaching history.

Instructional Design as Inquiry

At its core, instructional design is an act of inquiry. History educators begin with a blank page and their own content knowledge, resources, and standards. They wonder, "What do I want students to understand? How will they show it? What learning experiences will get them there?"

The problem they presently face is that GenAI can populate a lesson plan instantly. We have seen it! Yet, when examined, the abundance of words and ideas often conceals emptiness or focus on recall (Trust & Maloy, 2025). It is just AI slop. The challenge is no longer generating content for a lesson but deciding what matters. It is the same issue faced by history educators trying to discern high-quality online educational resources from low-quality resources. How do you discern?

If Lovelace offers a philosophy of design, she also offers a warning. The machine, she knew, could only ever reflect the assumptions embedded within it. "The Analytical Engine," she wrote, "weaves algebraic patterns just as the Jacquard loom weaves flowers and leaves" (1843). A loom can reproduce any pattern the designer programs, but it cannot question the pattern itself. That responsibility belongs to the human artisan.

The same is true of Chatty Geeps. It can generate lesson plans, prompts, and assessments that appear persuasive, but its logic is inherited rather than original. It draws from the woven patterns of the internet, including its biases, omissions, and blind spots. The teacher's role, then, extends beyond prompting the machine. It also involves examining the pattern it produces and deciding what needs to be reworked. From a technocurious approach, we can treat GenAI as a drafting partner rather than a deliverer or doer. Chatty Geeps can join the inquiry to assist in designing questions and to help extend the questioning. It can both mirror and multiply teacher thinking when we push it, stress test it, and speculate. To do this, we might enter prompts like:

- *Generate three possible compelling questions aligned with C3 Dimension 1 for this state standard.*
- *Suggest how this objective might be represented in a performance task rather than a worksheet.*
- *Offer three ways to scaffold this inquiry for multilingual learners without reducing cognitive demand.*

In our initial draft of this book, we did not include a chapter on instructional design or lesson planning. Admittedly, it's not the first thing we did when we started working with Chatty Geeps. But it is also the application that we have seen teachers go to repeatedly. And that makes sense! Planning is where the demands of teaching collide most directly. It's tempting to hand the work to the machine and call it efficiency. But when planning with Chatty Geeps, the question should not be "Can GenAI plan for me?" but "How can GenAI think with me?"

Designing for Understanding and Meaningful Work

History teachers have long wrestled with the balance between *coverage* and *understanding*. Two frameworks help teachers navigate these tensions:— Understanding by Design (Wiggins & McTighe, 2005) and Authentic Intellectual Work (Newmann et al., 2001; Newmann et al., 2007; Newmann et al., 2016). Both are important parts of how we design for student learning.

Understanding by Design (UbD) gives teachers a map for planning instruction. Its central idea is simple but transformative: *start with the end in mind*. Instead of beginning with activities, teachers first define what enduring understandings they want students to walk away with, then they plan backward to align assessment and instruction. The three stages of UbD are:

1. Identify desired results:— What big ideas, understandings, and essential questions anchor this unit?
2. Determine acceptable evidence:— How will students show what they understand?
3. Plan learning experiences and instruction:— What strategies, activities, scaffolds, and supports will lead my students there?

In history education, UbD ensures that inquiry is deliberate. It links standards to compelling questions and provides a through line from objectives to assessment. UbD helps us resist the urge to simply "cover" material and instead encourages designing for transfer, moving historical understanding into new contexts.

We also recognize that one of the challenges of being a history teacher is convincing students that what we are teaching matters and is relevant, situating learning history as more than remembering facts for a trivia quiz on Tuesdays at the pub, passing the end-of-course multiple-choice test, or hunting for the right word in slot notes.

Authentic Intellectual Work (AIW) is a powerful nudge toward stressing history's relevance. AIW shifts attention from *what we plan* to *the quality of what students actually do and why it matters.* It offers three simple but powerful criteria:

1. Construction of knowledge:— Students actively build and interpret meaning rather than recall facts.
2. Disciplined inquiry:— Students use evidence, reasoning, and communication grounded in the disciplines.
3. Value beyond school:— The work connects to real audiences and problems that matter beyond the classroom.

In history education, these criteria push lessons toward intellectual and human relevance or significance. Instead of a prompt like *Write three facts about Jamestown*, students might ask, *How did geography shape early Virginia settlements?* and present their findings to a community audience.

Together, UbD and AIW support inquiry-based, standards-aligned, and meaningful learning. These frameworks also align with the technocurious stance that runs through this book. Chatty Geeps can populate a unit plan in seconds, but it cannot decide what is worth understanding or why that understanding matters. That's still the work of the teacher. By pairing UbD's structure with AIW's depth, teachers can use GenAI as a tool for design, not as a substitute for thinking.

Pairing UbD with the Centaur Model

Across our work, we've come to describe AI-supported instructional design through what Ethan Mollick (2024) calls the Centaur model, referring to the partnership between human discernment and GenAI speed. Like the mythic creature itself, this approach combines multiple intelligences: human creativity and contextual understanding on one hand and machine pattern recognition, recall, and draft generation on the other.

For history educators, the Centaur model offers a way to visualize shared labor between teacher and tool. It is a hybrid form of planning that keeps human judgment at the center. Teachers provide the discernment through their contextual knowledge of students, community, ethics, and standards. Chatty Geeps contributes the computational power, scanning, aligning, organizing, and suggesting. We can pair this model with the C3 framework, state standards and the structure of UbD:

Stage 1 (Identify Desired Results): Chatty Geeps can help teachers ensure enduring understanding, reword standards into compelling questions, or check alignment with frameworks like the C3 framework or Virginia Standards of Learning. The teacher's task is to choose what matters and decide which understandings are worth pursuing at length.

Stage 2 (Determine Acceptable Evidence): Chatty Geeps can brainstorm authentic performance tasks or rubrics, suggest ways to assess disciplinary thinking, or generate sample responses for discussion.

Stage 3 (Plan Learning Experiences and Instruction): Chatty Geeps can help sequence lessons, suggest scaffolds, or differentiate resources for multilingual learners. Yet even here, the teacher remains the curator, deciding which experiences will help students think critically about evidence and connect learning to civic life.

The chart below demonstrate what this might look like in practice. Chatty Geeps may quicken the drafting process, but only thoughtful teachers can ensure that what's designed is authentic, rigorous, and human. Experimenting and playing with new tools while holding fast to our pedagogical compass is the heart of technocuriosity.

Phase	Teacher Discernment	GenAI Contribution	Centaur Collaboration
Visioning UbD	Identifying big creative ideas for inquiry, disciplinary goals, and purposes; connecting to learners' contexts; defining disciplinary purpose	Revealing thematic connections across standards and content; proposing enduring understandings; suggesting disciplinary-specific compelling questions; mapping standards	Generating drafts of enduring understandings and essential disciplinary questions for critique and refinement AI drafts → teacher critiques → shared revision
Designing UbD	Creating authentic tasks; providing disciplinary ideas and strategies; anticipating misconceptions; considering access for all students	Suggesting example tasks/activities alignment with different types of assessment formats aligned with rubric criteria; refining scaffolds from exemplars	Codesigning scaffolds and performance assessments; refining rubrics for clarity and bias or oversimplification
Building UbD	Knowing students; adjusting pacing; embedding scaffolds, discussion, and disciplinary supports; sequencing experiences to support the needs of students	Drafting lesson sequences; sequencing lessons; offering resource suggestions; modeling think-alouds for disciplinary thinking; flagging missing scaffold or resources	Merging GenAI pacing and models with classroom reality, real student readiness, and classroom rhythms, adjusting for time, tools, and students

Phase	Teacher Discernment	GenAI Contribution	Centaur Collaboration
Reflecting/ Testing AIW/UbD/ C3 Alignments	Evaluating intellectual quality, rigor, student agency, access, and coherence	Generating analytic summaries of plans; analyzing crosswalking standards; testing alignment between goals, assessment, and strategies	Using Chatty Geeps for reflection prompts (e.g., asking *Which parts of this plan promote construction of knowledge?* then recording output and responding critically)

Figure 6.2: Understanding by Design in collaboration with Chatty Geeps

The Centaur model is less a division of labor than a conversation between capacities. Teachers contribute judgment, empathy, and an understanding of what matters for students in their context. Chatty Geeps contributes breadth, including the capacity to scan, sort, and pattern across data. Together, they codesign for both depth and reach.

For example, a teacher working on a world history inquiry about imperialism might begin by brainstorming essential ideas:

> *Generate three possible enduring understandings about the legacies of imperialism that align with C3 framework Dimension 2.*

Chatty Geeps might reply with:

> *Empires reshape global systems of power, trade, and culture in ways that continue to influence today's world.*
>
> *Imperial narratives often privilege one perspective while silencing others.*
>
> *The process of decolonization reshaped notions of sovereignty and national identity.*

From here, the teacher acts as a critic. They interrogate the phrasing, align it to standards, and ensure historical accuracy. The tool provides the pattern, the teacher provides the discipline, and together, they refine, clarify, and reframe. Chatty Geeps is a remarkable accelerator for drafting, but it is the teacher's disciplinary expertise that ensures integrity.

GenAI and Understanding by Design in Practice

Backward design provides the structure. The Centaur model helps distribute the cognitive load. But what does this actually look like in practice? The following "technocurious map" shows how Chatty Geeps can collaborate across each stage of the Understanding by Design framework while keeping human discernment at the helm.

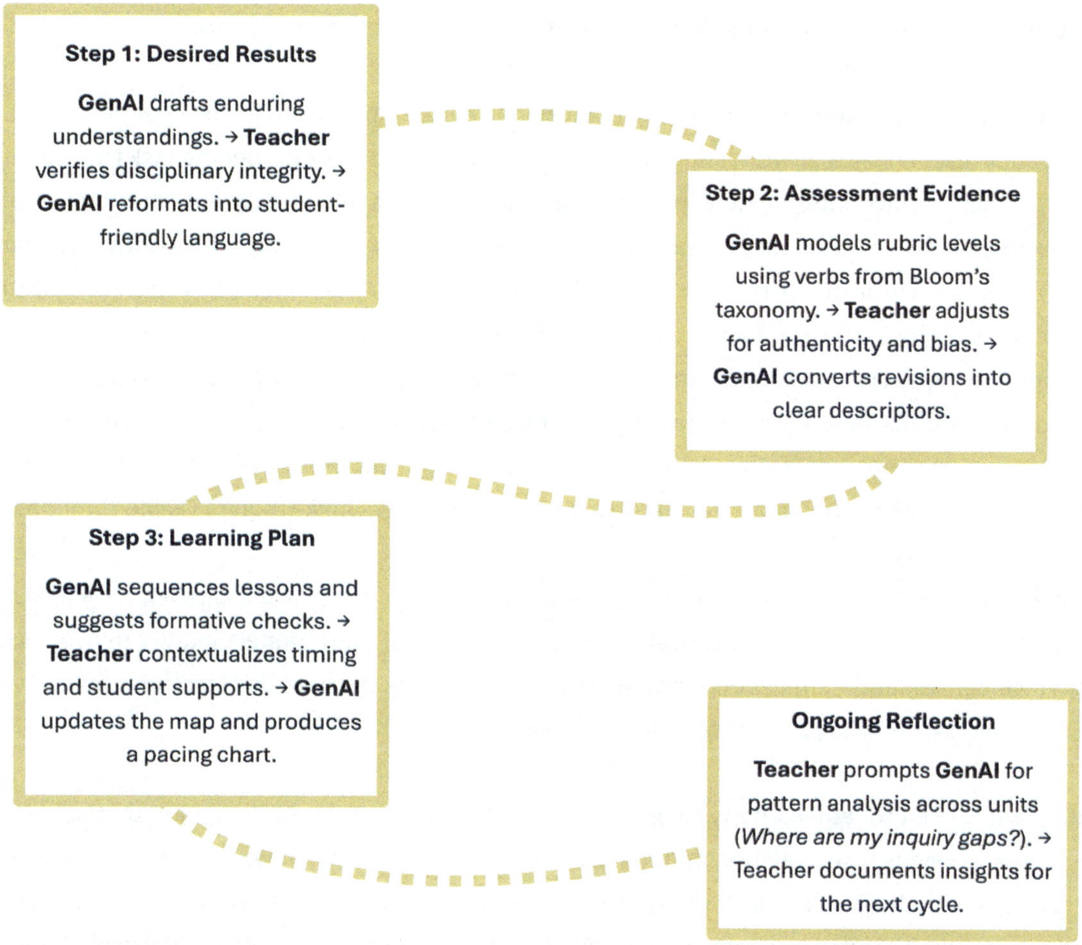

Figure 6.3: AI in UbD stages: A technocurious map

In this model, GenAI becomes a mirror, not a map. It reflects our designs back to us, revealing patterns we might not notice, but it does not tell us where to go. The backward design process becomes a feedback loop where each stage ends not with acceptance but with amendment. It is an iterative conversation between anticipation and revision, precision and play.

Integrating the Three Criteria of Authentic Intellectual Work

If backward design provides the structure, then a framework like Authentic Intellectual Work (AIW) becomes the quality filter. Does the task require students to construct knowledge, engage in disciplined inquiry, and produce something of value beyond school? If not, it's back to the drawing board.

When paired with Chatty Geeps, we might think about the three tenets of AIW like this:

1. **Construction of Knowledge:** Chatty Geeps may provide Bloom's-based, higher level, actionable verbs that everybody loves (*analyze, connect, evaluate, argue*), but it cannot judge whether the task truly requires synthesis. Teachers must interrogate whether the activity demands intellectual work or merely rephrasing. For example, Chatty Geeps may propose the exercise *Summarize the causes of the French Revolution*, but a teacher can push it further, arriving at *Use sources from different estates to evaluate competing explanations for the Revolution*. The tool offers fluency while the teacher ensures rigor.
2. **Disciplined Inquiry:** Custom GPTs can be trained to reference disciplinary tools like sourcing, contextualization, and corroboration. Yet only teachers can model these moves authentically. For example, a model can generate the sentence frame: *The author's perspective might have influenced…*, but it cannot replicate the embodied work of reasoning with evidence in real time. Teachers make inquiry visible through talk, gesture, and improvisation.
3. **Value Beyond School:** Chatty Geeps can gesture toward relevance (*Connect the New Deal to modern economic debates*), but it cannot locate local meaning. Only teachers can connect inquiry to lived experience that includes students' families and communities. Chatty Geeps can help find local archives or civic issues, but it is the teacher who gives those connections consequence.

Ultimately, when used together, Chatty Geeps, UbD, and AIW form a self-checking loop that asks "Does this design build knowledge, use evidence, and matter in the world?" And when we align the prompts provided to Chatty Geeps with the AIW framework, we find the boundaries of its usefulness. Chatty Geeps can scaffold thinking; it cannot substitute for judgment. It can propose tasks that look authentic, but teachers must ensure that they feel authentic and that they matter to learners as citizens, not just as test-takers.

Critical AI Literacy in Lesson Design

We've said this before and will say it again: critical AI literacy is not about technical fluency. It's about critical discernment.

To be technocurious is to understand both how GenAI works and how to interrogate its workings. It's about developing habits of reading and annotating GenAI outputs with the same scrutiny we apply to historical sources. Just as Lovelace wrote notes in the margins of Babbage's designs, teachers annotate, question, and revise the work of their algorithmic collaborators.

Four Habits of Critical AI Literacy for Technocurious Designers

- *Attribution: Always know the source. Ask "What data might this response draw from?" and note what remains invisible.*
- *Alignment Verification: Ask Chatty Geeps to crosswalk standards, then confirm manually.*
- *Bias Check: Pose counterfactuals. For example, "Would this task work equally well for a student from a different historical or linguistic background?"*
- *Reflection Loop: Keep a design log noting what Chatty Geeps contributed, what you changed, and why.*

In our own speculative work, we've learned that the most valuable outputs are not the ones that get it right but the ones that get it almost right.

Designing for Difference: Scaffolding and Student Diversity

Strong instructional design must account for the realities of classrooms that include students with different languages, literacies, motivations, and experiences. Chatty Geeps can serve as a collaborator in this complexity, helping teachers imagine scaffolds, alternatives, and pathways for access that might otherwise take hours to draft manually. In this role, Chatty Geeps is less an automator than a co-teacher, serving as a planning aide who can draft supports, visualize adaptations, and propose differentiation options for reflection and refinement.

In our own work, Chatty Geeps has helped prototype scaffolds, anticipate misconceptions, and design multiple pathways for expression and understanding. We continue to play and test how Chatty Geeps can adapt/level our historical sources and provide tiered vocabulary and support. We have asked it to codevelop alternative assessments to demonstrate understanding through choice, including audio reflection, diagrams, or short written briefs. We used it to anticipate alternate concepts students may come to when discussing historical topics (e.g., *What alternate concepts or misconceptions might students make when analyzing the causes of … ? What mistakes might a student make when analyzing this source?*). GenAI can help visualize scaffolds and supports, especially for multilingual learners or students with varying readiness. For example, Chatty Geeps can:

- generate tiered vocabulary lists from a historical source or simplify historical sources;
- suggest gradual release strategies for inquiry-based lessons; and
- provide alternative modalities (timeline, graphic organizer, short summary).

Example Prompting Moves

- *Based on this lesson plan, identify where multilingual learners might need language scaffolds.*
- *Suggest inquiry adaptations for students with limited background knowledge.*
- *Simplify the language of this source so it is accessible for a student reading at a second-grade level.*
- *How can I do a slow reveal of this historical source for a ninth-grade class?*

The outputs can be uneven, but the process itself builds teacher sensitivity. It helps us see where our lessons assume too much, move too fast, or rely on hidden knowledge. Designing for difference is about flexibility. With this in mind, Chatty Geeps can suggest access points as beginning places for teachers to consider.

Choosing Your Function

Not all design work needs to happen in the same space. Each of Chatty Geeps's modes (chat, custom GPTs, projects) serves a different purpose, tempo, and degree of teacher mediation. The technocurious stance means experimenting with each deliberately, recognizing that every format structures thinking in different ways.

- **Regular Chat:** Chat mode is best for rapid brainstorming, idea generation, or first drafts. Teachers might use it to explore possibilities, e.g., *Give me five possible ways to connect Reconstruction to contemporary debates about democracy*. The responses may range from insightful to absurd. That's the point. Chatty Geeps becomes a sounding board, a way to see your own ideas refracted back through the algorithmic mirror.
- **Custom GPTs:** Custom GPTs are trained with specific frameworks, resources, or district standards. These can be built as "design partners" tuned to UbD, AIW, or C3 frameworks. For instance: *Use the Instructional Design Mentor to check whether this formative task aligns with my objectives and C3 Dimension 2*. Custom GPTs extend a teacher's cognitive reach without displacing judgment. They scaffold alignment, but the teacher still decides what counts as meaningful learning. For more about building and working with custom GPTs, check out Chapter 8.
- **Projects**: Projects are best for iterative design cycles like unit collections, rubrics, or curriculum mapping, where teachers need a persistent workspace for collaboration. Within a world history inquiry project, for example, a teacher might type *Compare the rubric for Lesson 3 to Lessons 1 and 2. Are the performance expectations aligned?* Chatty Geeps will generate an analytic summary or alignment map; the teacher evaluates, revises, and decides what to keep.

In practice, these modes blur together. We do not see them as fixed categories but as a set of evolving affordances that are like different doors in the same hallway, opening to different kinds of thinking. Some, like custom GPTs and projects, show direct promise for supporting disciplinary design and reflection. Others that we haven't mentioned here, like deep research or agent modes, remain experimental and require heavy mediation. It's not about mastering every mode or investigating every door in the hallway. The key is choosing intentionally.

Through the Looking Glass: Six Impossible Things Before Planning

Before your next planning session, try practicing a bit of impossible thinking.

Backward design asks us to picture evidence of understanding before it exists. Technocurious design asks us to imagine how GenAI might help us get there without losing sight of what only humans can do.

Use this reflection to stretch your planning imagination:

The Impossible Scaffold: *What if GenAI could model curiosity rather than completion? How would that change the way you design prompts or inquiry arcs?*

The Impossible Mirror: *What if your unit plan could show you what you've overlooked, whose voices, perspectives, or misconceptions are missing?*

The Impossible Student: *What if your quietest learner found their voice through an AI-assisted conversation about evidence or perspective?*

The Impossible Alignment: *What if backward design felt less like compliance and more like creative composition, an act of crafting meaning rather than mapping standards?*

The Impossible Feedback Loop: *What if GenAI could illuminate your teaching patterns, not to automate them but to help you see your own habits of mind?*

The Impossible Future: *What if teaching with GenAI became less about managing a machine and more about imagining what learning could be?*

Reflection Prompt:

- Which of these six impossible things feels most possible in your current practice?
- Which still feels out of reach, and what small experiment could bring it closer?

Moving Forward with Imagination and the Impossible

Ada Lovelace once said that "imagination ... is that which penetrates into the unseen worlds around us" (Ignotofsky, 2016). For her, the great promise of the analytical engine was not computation; it was creativity. Similarly, the promise of Chatty Geeps in lesson planning may not be in the probabilistic predictions it can make or the speed at which it can generate a multiple-choice quiz, but in the ways it can act as a creative thinking partner. It can mirror our reasoning, challenge our assumptions, and sometimes even surprise us. When used well, Chatty Geeps doesn't erase the teacher; it amplifies the teacher's imagination. For educators, the unseen world is not made of numbers or gears but of possibility, of new ways to connect students to history, to complexity, and to each other.

And this may be the space in which Ada Lovelace and the White Queen overlap in their thinking. What Ada called imagination, the White Queen describes as believing in the impossible. When Alice protests that she "can't believe impossible things," the Queen gently replies, "I daresay you haven't had much practice. When I was your age, I always did it for half an hour a day. Why, sometimes I've believed as many as six impossible things before breakfast."

The kind of pedagogical imagination that GenAI makes possible may allow teachers to envision evidence of learning before it exists, to anticipate understanding that hasn't yet taken shape, to believe that a tool built for text generation could become a scaffold for inquiry. Rather than positioning students as historians, the aim is to help them take up the habits of historical thinking like posing questions, weighing evidence, and constructing interpretations. In this sense, generative AI can support students in developing their own interpretive voices and reasoning processes, not by imitating expert historians but by practicing the discipline in ways that are authentic to their level of learning.

But this work demands intention rather than convenience. Using Chatty Geeps to generate lessons because we're tired or pressed for time holds back both the tool and the profession. History teaching is more than fill-in-the-blank exercises. It's a craft of judgment, empathy, and design. When we hand that craft over to automation, we trade imagination for imitation. Instead, the real opportunity of Chatty Geeps lies in using it *with* us, not *for* us, refining questions, prototyping scaffolds, exploring alternatives, and testing alignment. It can help us plan more ambitiously rather than just more quickly. It can turn the planning process itself into an inquiry space where teachers practice curiosity, reflection, and critique.

Backward design, like the White Queen's way of living, asks us to look forward into the future and imagine what learning might look like before it happens. It is, in its own way, living backward:— a practice of seeing understanding before it's visible, of designing evidence before it emerges. And like Alice, we may find that process dizzying at first, full of contradictions and nonsense. But in that very disorientation lies creative power. If we can approach GenAI with that same sense of disciplined wonder, we just might find that what once looked like a shortcut machine becomes a companion for thinking, planning, and reimagining the future of history education itself.

Through the Looking Glass: Designing with Intention

- *When Chatty Geeps can generate lessons in seconds, what does it mean to design instruction with care and purpose?*
- *How might teachers use GenAI to scaffold learning without surrendering the craft of pedagogy?*
- *In what ways can thoughtful design keep human judgment, creativity, and context at the center of AI-assisted teaching?*

Image Long Description

Figure 6.3: Step 1: Desired Results. GenAI drafts enduring understandings. → Teacher verifies disciplinary integrity. → GenAI reformats into student-friendly language. Step 2: Assessment Evidence. GenAI models rubric levels using verbs from Bloom's taxonomy. → Teacher adjusts for authenticity and bias. → GenAI converts revisions into clear descriptors. Step 3: Learning Plan. GenAI sequences lessons and suggests formative checks. → Teacher contextualizes timing and student supports. → GenAI updates the map and produces a pacing chart. Ongoing Reflection. Teacher prompts GenAI for pattern analysis across units (Where are my inquiry gaps?). → Teacher documents insights for the next cycle.

7. GENAI AS THOUGHT PARTNER: SUPPORTING SELF-REGULATED LEARNERS WITH QFT

On a winter morning in 1823, Mary Anning crouched at the base of the cliffs of Lyme Regis. Before her lay an elongated neck, ribbed torso, and paddle-like limbs: the fossilized remains of a creature that no human eye had ever seen whole. She began to carefully and deliberately draw what she saw, sketching bone to bone in a journal entry that would later help paleontologists reconstruct the first known *Plesiosaurus*.[1] That sketch, now preserved in the archives of the Oxford University Museum of Natural History, captures the moment when evidence and imagination met on the edge of the known world (Eylott, 2018).

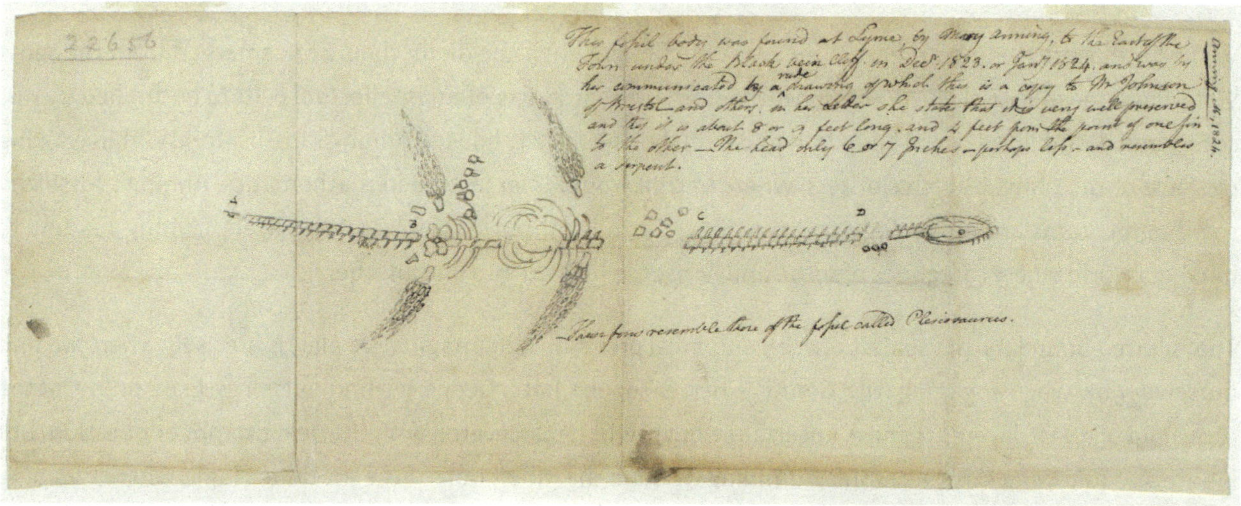

Figure 7.1: Pen and ink drawing of Plesiosaurus

Anning's work fundamentally changed how Victorians understood time itself. The fossils she uncovered and sold to collectors and scholars revealed a prehistoric world that predated the biblical flood narrative and hinted at extinction, an idea many of her contemporaries found theologically and emotionally unsettling. In letters exchanged among the learned societies of London and Oxford, Anning's finds were discussed with a mixture of

1. Little-known fact: although it was considered, the Plesiosaurus was not featured in any of the films in the Jurassic Park movie franchise.

Also, if you are here looking for the posters mentioned on page xiii, an editorial decision prevented their inclusion.

awe and disbelief. Anning's steady hand and empirical patience gave form to a past that both fascinated and frightened her peers.

Four decades later, Lewis Carroll sent another kind of explorer plummeting into the unknown:

> In another moment down went Alice after [the rabbit], never once considering how in the world she was to get out again. The rabbit-hole went straight on like a tunnel for some way, and then dipped suddenly down, so suddenly that Alice had not a moment to think about stopping herself before she found herself falling down a very deep well.

As she falls, Alice fills the air with questions: "Do cats eat bats? Do bats eat cats? ... I wonder how many miles I've fallen by this time?" In that long descent, reason and nonsense intertwine. She catalogues what she knows, tests possibilities aloud, and revises mid-thought. Her fall (almost) becomes a comic enactment of self-regulated inquiry.

While Carroll and Anning never met (he was only fifteen when she died), their work emerged from the same Victorian atmosphere of epistemic upheaval. Both were products of an intellectual culture enthralled by discovery and destabilized by it. The Oxford museum that eventually housed Anning's fossils stood within walking distance of the Christ Church cloisters, where Carroll would later lecture in mathematics. Anning's fossilized sea dragons transformed geology and theology; Carroll's Alice transformed logic itself, inviting readers to inhabit a world where categories of sense and nonsense blurred into one another.

That blurred boundary, the space between empirical precision and imaginative play, is precisely where we find ourselves now when engaging with GenAI. When we open Chatty Geeps, we find ourselves, like Anning, at the threshold of a vast, layered archive, uncertain what might be uncovered with the next prompt or question. But we are also following Alice's example, tumbling through the rabbit hole of the machine's logic, where meaning morphs with every iteration. Each response invites us into a technocurious mindset through both fascination and skepticism. It's normal to ask yourself, "Is what I'm seeing real, useful, true? Or is it simply well formed?"

For history educators, this parallel is instructive. What both Anning and Alice reveal, in their very different ways, is that learning is not a matter of accumulating facts but of learning how to learn. Anning reacting to a relic or source with steady sketching reflects the forethought of inquiry, as she considers where to dig, what to observe, what to record. Alice's tumbling curiosity captures a sense of performance that allows her to ask, test, and revise in motion. And both embody reflection when they pause to make sense of what they've seen. These cycles of planning, doing, and revising are the essence of self-regulated learning. In the history classroom, this process is made visible each time a student frames a historical question, confronts contradictory evidence, or revises an interpretation. In the pages ahead, we explore how GenAI can aid the recursive process of inquiry when used thoughtfully.

What It Means to Learn How to Learn

Instructional design is more than sequencing lessons or aligning standards. As teachers, we often take for granted how we learn, but learning involves creating the conditions for students to learn how to manage their own learning through explicit scaffolding. Self-regulated learners set goals, monitor progress, choose strategies, and reflect on results. In history classrooms, these activities mean learning to question sources, evaluate evidence, and trace how their thinking changes over time (Journell et al., forthcoming). Self-regulated learning is about becoming aware of what one is doing to support one's own learning, an act of metacognitive attention that mirrors how disciplinary thinking is carried out. If we can get students aware of what history is and how to do history at varying levels of progression, we will have achieved a nearly impossible feat.

Designing for self-regulated learning asks teachers to think beyond coverage or pacing and toward cultivating habits of mind. A well-constructed prompt, question, or reflection task can help students move through the recursive cycle that psychologists describe as forethought → performance → reflection, a process that closely parallels the logic of historical inquiry.

Forethought: Framing a Historical Problem

Forethought is the anticipatory stage of learning. This stage is when learners set goals, draw on prior knowledge, and define what success might look like. In history classrooms, this often takes the form of posing a meaningful question: What caused this event? How did people in this moment understand their world? Why did this happen? How could this have happened?

Chatty Geeps can support this stage by helping students articulate and refine their intentions. A prompt such as *Generate three versions of my research question about Reconstruction, moving from broad to specific*, encourages learners to see inquiry as iterative. When teachers design activities that model this forethought, students learn that good history begins not with an answer but with a problem worth pursuing.

Performance: Gathering and Testing Evidence

In the performance phase, learners enact their plans, reading sources, testing hypotheses, revising interpretations, and monitoring comprehension. It is the cognitive equivalent of Anning's excavation or Alice's mid-fall questions. Chatty Geeps can serve as a space for this kind of metacognitive dialogue if we learn to prompt for it deliberately and teach students to do the same. A simple request like *Provide three questions my students could use to monitor their understanding after analyzing a historical source* turns the tool into a mirror for reasoning. Custom

GPTs can scaffold these processes by posing follow-up questions (*What evidence best supports your claim?*, *What counterarguments might challenge it?*), reminding learners that inquiry is recursive rather than linear.

The goal is to design prompts that help students externalize their thinking. Ideally, these prompts help students to test, refine, and reconfigure their understanding as they go.

Reflection: Interpreting, Revising, and Re-Questioning

Reflection closes and reopens the cycle. It is the pause that allows students to evaluate what they've learned, where their reasoning holds, and where it falters. In historical inquiry, reflection might mean reconsidering which evidence was most convincing or how a single source reshaped their interpretation.

Imagine a student researching Reconstruction-era schooling. Initially, she frames her question around access to education. Halfway through, she discovers Freedmen's Bureau reports that complicate her assumption, leading her to revise her question: How did newly freed communities define education for themselves? Her ability to adapt and reframe the question in light of new evidence is self-regulated learning in motion.

Chatty Geeps can reinforce this stage with small, habitual prompts: *What evidence most changed your view?* or *Where do you still feel uncertain?* Reflection becomes not an afterthought but an integrated rhythm of learning.

Linking Self-Regulation and Instructional Design

Designing for self-regulation involves several deliberate moves:

- ***Goal setting.*** *Make objectives transparent and revisit them throughout a unit. Encourage students to predict what success might look like.*
- ***Monitoring.*** *Embed small formative checks that ask students to pause and evaluate their understanding before feedback arrives.*
- ***Reflection.*** *Close lessons with short prompts like* What surprised you most? *or* What might you still need to investigate? *so reflection becomes routine rather than episodic.*

Chatty Geeps can prompt teachers to build these opportunities into their plans. A custom GPT can suggest metacognitive questions, generate reflective sentence starters, or translate success criteria into student-friendly language. Over time, these digital traces, including student reflections, evolving prompts, and self-assessments, become a record of metacognitive growth.

Yet discernment still belongs to the teacher, who decides which reflective habits matter most for their students. The centaur model applies here. Chatty Geeps can scaffold regulation by offering prompts, patterns, or checkpoints, but the stu-

dent must steer the reflection. The goal of design is agency, supporting learners who can think historically, reason with evidence, and recognize how they learn.

Through the Looking Glass: Mirror Moment

- *What visible signs of self-regulation do you see in your students? How often do they pause to reconsider their own reasoning? How often are they prompted to pause? What routines could make this reflection more visible?*

The Question Formulation Technique as Inquiry Engine

In *Alice's Adventures in Wonderland*, curiosity propels the story forward. The Question Formulation Technique (QFT), developed by the Right Question Institute (https://rightquestion.org) (2025), offers teachers a way to bring that same disciplined curiosity into the classroom. It is both simple and radical, laying out a step-by-step method for teaching students not what to ask, but *how* to ask.

The QFT unfolds in a short, structured sequence that can be implemented in a variety of ways. Students begin with a Question Focus (QFocus), which is a statement, image, quotation, or artifact that serves as a spark. Working alone or in small groups, they produce as many questions as they can, following specific rules: ask questions, do not judge, record every question exactly as stated, and convert any statements into questions. Then, they improve their questions by labeling each as *open* or *closed* and determining how to open the closed questions, and vice versa. Next, they prioritize. They select three questions that matter most moving forward. Finally, students plan how to pursue those questions and reflect on what they have learned about both content and process. Within these few steps (produce, improve, prioritize, and reflect), students enact the full cycle of self-regulated learning.

What makes the QFT powerful is not the elegance of its design but the shift it enables. Rather than answering the teacher's questions, students generate, classify, and revise their own. Ownership of inquiry moves from teacher to learner. This shift demands forethought ("What do I want to know?"), performance ("How can I phrase this clearly?"), and reflection ("Which questions are worth pursuing and why?"). As students listen to one another's language and refine their phrasing, they become aware of their own reasoning processes. They define problems, test interpretations, and discern which questions make the most sense for where they are headed next.

Yet the QFT is not without challenges. Left unscaffolded, students often default to recall-based questions like "When was this photo taken?" or "Who is in it?" that skim the surface rather than probe meaning. Others struggle to revise or rank their questions and are reluctant to let any go. These are natural stages of metacognitive growth. When teachers name these moments explicitly ("Notice how many of your questions are about what happened versus why."), students begin to monitor and adjust their own inquiry habits. We are not working towards a perfect list of questions but a conscious process for generating and evaluating them.

For history teachers, the QFT can serve as the engine of inquiry rather than a pre-lesson warm-up. A well-chosen QFocus, like a political cartoon, a Freedmen's Bureau report, or a photograph from the Great Migration, can generate an entire unit's worth of investigation. Students' priority questions become the scaffolds for research, writing, and discussion. In this way, the QFT rehearses the work of historical thinking, noticing complexity, revising initial assumptions, and tracing cause and consequence through evidence.

Critics of the QFT often ask, "What do you do with it next?" or frame it as a waste of time. However, we find that the questions generated by students often align with the standards included in a unit plan. As you move through the unit or lesson, the questions can be posted so students can check off what they have answered. Ideally, these questions are not just tangents but can be incorporated into the inquiries that teachers have already planned or used to facilitate discussions working toward synthesis. To treat this as a one-and-done activity potentially harms student voices rather than amplifying them.

Chatty Geeps as Question Coach

If self-regulated learning begins with curiosity, then one of the teacher's most important roles is to help students sustain that curiosity long enough for it to do real intellectual work. Chatty Geeps, when used wisely, can serve as a question coach. The key is in the framing. Chatty Geeps should support the asking, not shortcut the answering.

When teachers introduce Chatty Geeps as a thought partner, they invite students to externalize the earliest stages of historical thinking when a question is only half-formed. As outlined in the example below, in this role, Chatty Geeps helps learners practice inquiry routines, moving through the steps that parallel the Question Formulation Technique of generating, improving, and prioritizing questions.

Step 1: Question Focus

A seventh-grade class is presented with a photo of Manzanar, a Japanese incarceration camp in California (although they don't yet have that context).

Figure 7.2: Baseball game, Manzanar Relocation Center, Calif.

Stage 2: Generating Questions

When students approach a new topic, their questions often cluster around the factual (who, when, and where) before expanding toward the interpretive (why and how). Chatty Geeps can help stretch that range.

In small groups, they list as many questions as possible. After they complete the assignment, the teacher models what Chatty Geeps[2] can do. She shares the primary source image along with a prompt asking the model to create a list of questions about this source. Chatty Geeps then rapidly creates a list of possible questions.

2. When we have done this in real time, we ask these prompts in our custom History Mentor GPT, which has already been trained on what the QFT is. To access and use this custom GPT, visit here: https://chatgpt.com/g/g-D68jmCAro-history-mentor-for-core-concepts. For more information about how to create your own custom GPT, see Chapter 8.

History Mentor for Core Concepts 4o ⌄ Share ···

Absolutely — let's focus entirely on **Step 2 of the QFT**: *producing as many questions as possible* based on this primary source photo.

Here is a list of questions generated directly from observing the image. They are not judged, answered, or discussed — just listed as asked:

---

1. Why are so many people watching this baseball game?
2. Where is this game being played?
3. Who are the players on the field?
4. Are the people in the crowd part of a specific community?
5. When was this photograph taken?
6. Why are there mountains in the background — what does that say about the location?
7. What kind of uniforms are the players wearing?
8. Why is the field so barren and dusty?
9. What are the buildings in the background used for?
10. Are the people watching this game free to come and go as they please?
11. Why does the crowd seem to be standing behind a barrier?
12. What emotions might the crowd be feeling?
13. Is this a professional game or somethi ↓ nformal?

Figure 7.3: Custom GPT output for step two of the QFT

Once these questions have been generated, student groups can compare them with their own questions, asking:

- How are my questions the same or different from the questions created by Chatty Geeps?
- Are there questions from the AI-generated list that I would like to add to my group's list?
- Are there ways I can expand my list based on the ideas generated by this conversation?

At this juncture, the teacher may give students time to add to and expand their lists. When used this way, Chatty Geeps becomes a tool for divergent thinking. It models abundance rather than authority.

At this point, we might pause for a moment of self-reflection. Did Chatty Geeps help widen curiosity or deepen complexity? This is the core metric for AI-assisted questioning. If the output expands the scope of inquiry without dictating direction, it has done its job. As an extension, teachers could reinforce this stance by asking students to annotate AI-generated lists, identifying which questions feel promising, which feel irrelevant or biased, and which reveal gaps in their own understanding. This simple move preserves agency, and students remain in control of their own learning.

Stage 3: Improving Questions

Once students have a working list, Chatty Geeps can help them refine quality through metacognitive cues by asking questions about questions. Moving forward with the example from above, a teacher could choose one of two options.

In option one, individually or in small groups, students take pictures of their lists and upload them to Chatty Geeps along with a prompt (e.g., *Review my list of questions about this photo. Suggest two ways I could make them better, but don't rewrite them. Focus on clarity and depth*).

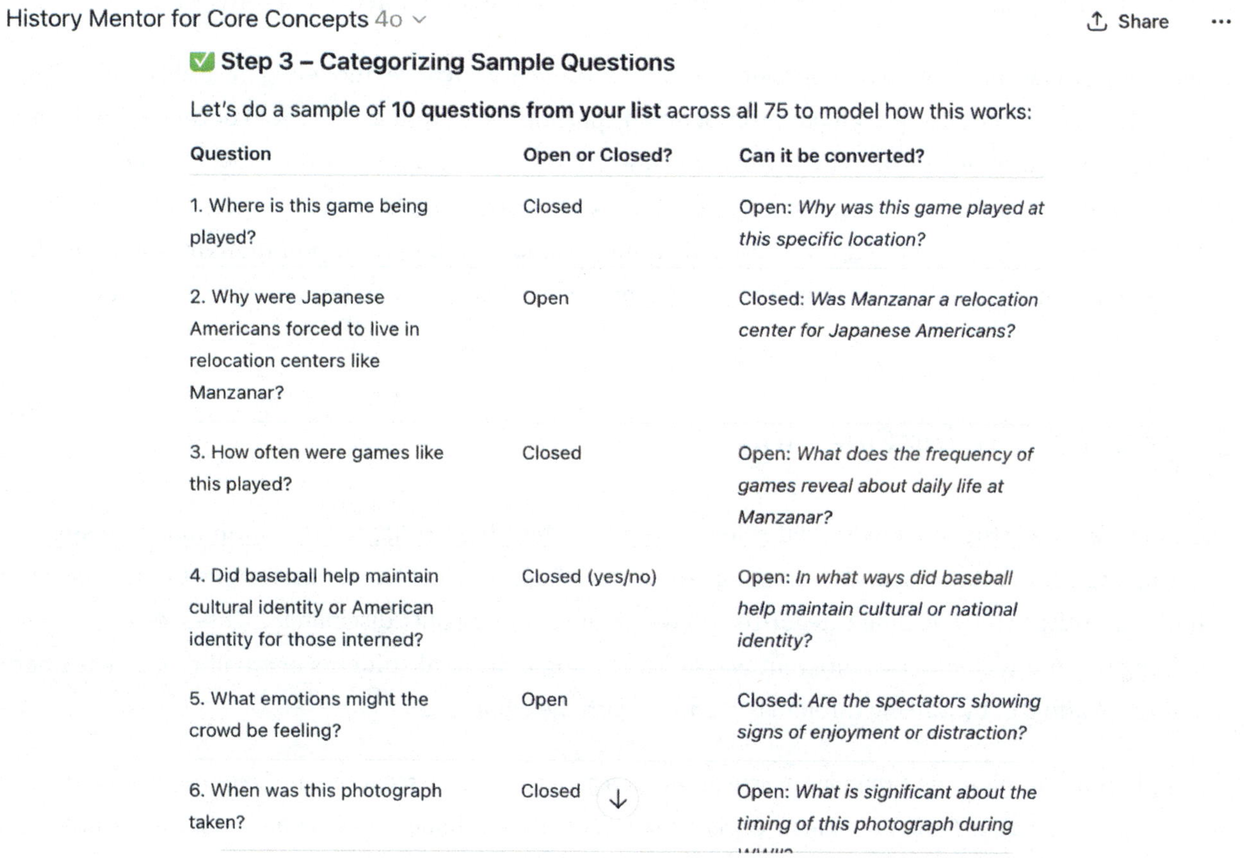
History Mentor for Core Concepts 4o ⌄ Share ···

✅ **Step 3 – Categorizing Sample Questions**

Let's do a sample of **10 questions from your list** across all 75 to model how this works:

Question	Open or Closed?	Can it be converted?
1. Where is this game being played?	Closed	Open: *Why was this game played at this specific location?*
2. Why were Japanese Americans forced to live in relocation centers like Manzanar?	Open	Closed: *Was Manzanar a relocation center for Japanese Americans?*
3. How often were games like this played?	Closed	Open: *What does the frequency of games reveal about daily life at Manzanar?*
4. Did baseball help maintain cultural identity or American identity for those interned?	Closed (yes/no)	Open: *In what ways did baseball help maintain cultural or national identity?*
5. What emotions might the crowd be feeling?	Open	Closed: *Are the spectators showing signs of enjoyment or distraction?*
6. When was this photograph taken?	Closed	Open: *What is significant about the timing of this photograph during*

Figure 7.4: Custom GPT output for step three of the QFT

Chatty Geeps might respond:

- *How might this question invite multiple viewpoints?*
- *Could this question be answered with a simple fact, or does it require interpretation?*
- *Does your question assume a single cause or allow for complexity?*

Here, the machine's strength lies in prompting reflection rather than revision.

Students could also ask the GPT to categorize each question as closed or open-ended and to model how to change these questions from one type to another.

Importantly, before students adopt any feedback from Chatty Geeps, they should critique it. Was the suggestion helpful, superficial, or misguided? Did it actually make the question stronger? This step reinforces authorship and helps students distinguish between mechanical correctness and intellectual correctness.

Alternatively, the second option for improving questions involves the teacher modeling the use of Chatty Geeps as a whole-class activity, using a similar prompt and displaying the output as a scaffold or think-aloud before asking students to return to their own questions and make improvements on their own. In modeling this process aloud, a teacher might ask questions like "Chatty Geeps says I should 'make this question more specific.' But sometimes specificity limits curiosity. What if I want to explore multiple interpretations instead?" This kind of metacommentary demonstrates that critical distance, not deference, is an important aspect of disciplined inquiry.

Stage 4: Prioritizing Questions

The final step asks students to sort and select. Too often, this phase collapses into intuition, with students choosing the questions that sound interesting rather than those that yield historical insight. Chatty Geeps can scaffold prioritization by helping categorize and weigh questions against disciplinary lenses. Again, teachers could engage in this step in two different ways:— by letting individual students or small groups use Chatty Geeps as a scaffold or by walking through the activity with the whole class.

A sample prompt might be *Group my questions into categories: cause, consequence, or significance. Suggest which might lead to the richest historical analysis* or *Sort these questions according to the six historical thinking concepts. Explain your reasoning*. By naming categories explicitly, Chatty Geeps externalizes a process that historians and history educators perform intuitively: sorting questions by type and potential yield.

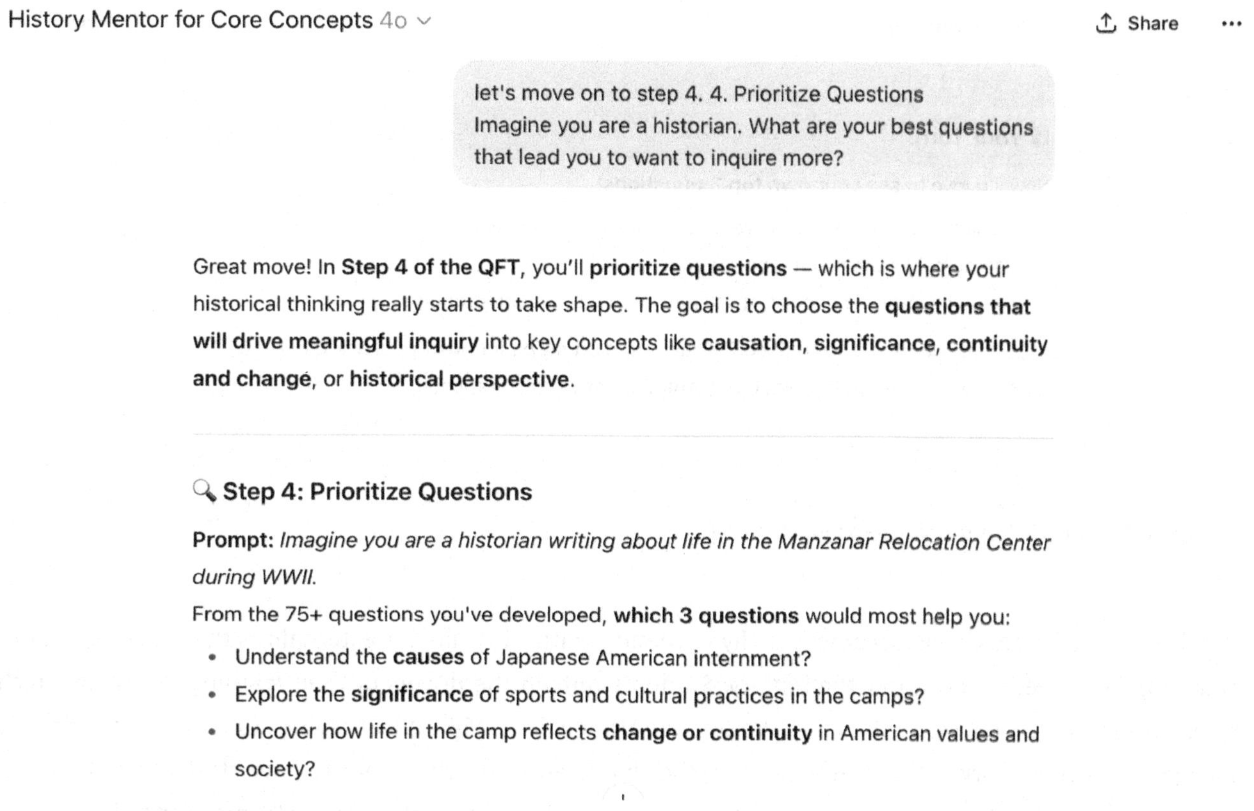

Figure 7.5: Custom GPT output for step four of the QFT

As a whole-class activity, implementing the QFT in this way may lead students to debate the rankings. One student may argue that "race and class" feels too broad while another insists that wartime context is the hinge. The ensuing discussion may reveal something that Chatty Geeps cannot comprehend: historians and history educators define significance differently depending on scale, evidence, and moral frame.

At this point in the activity, the teacher and students will choose questions they want to explore further throughout the lesson or unit. The teacher can intentionally select questions that already connect with state standards or that are forthcoming in the lesson they have already designed, but they can also supplement this list with student questions that allow for extension of the topic and continue to stimulate student interest.

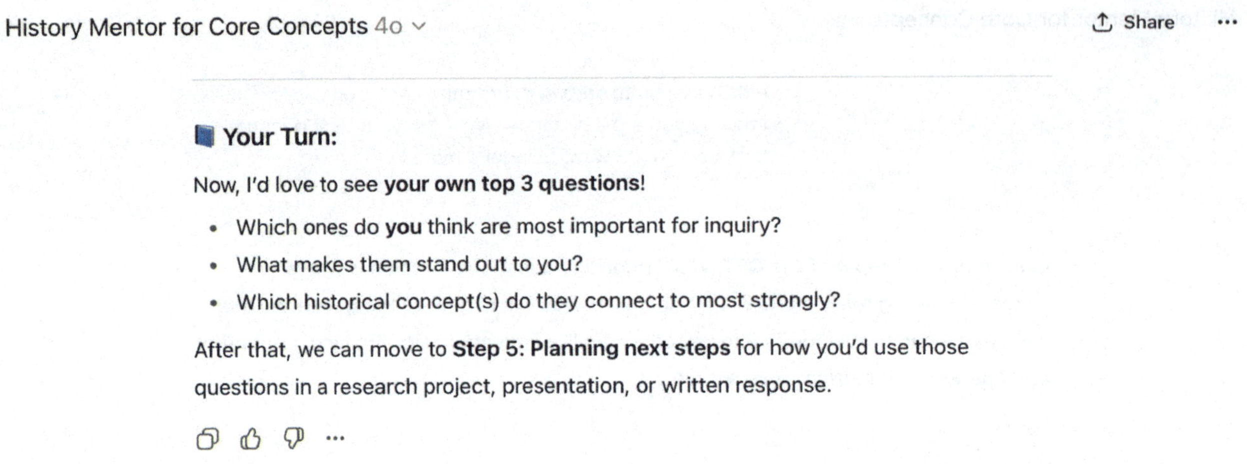

Figure 7.6: Custom GPT output for step four of the QFT

Used this way, Chatty Geeps becomes an ally in metacognitive learning. It externalizes the invisible work of thinking, but it never takes over the inquiry. Students remain the drivers of their learning, deciding which questions to pursue, revise, or abandon. We see GenAI's role in the QFT encouraging students to consider the navigation of their work without ever steering the ship itself. The point is not to have students write better questions faster but to see their own questioning as a process worth studying. When curiosity becomes a shared conversation between teacher, student, and machine, learning itself grows more visible.

MAD HATTER ALERT: Beware the Overeager Assistant

Even the best prompts can turn counterproductive when Chatty Geeps slips from coaching into answering. Teachers should watch for two warning signs:

1. ***When Chatty Geeps gives you the answer before you've even asked.*** *For example, a student types* Help me ask good questions about the Civil Rights Movement, *and the model responds with a mini-essay on Martin Luther King Jr. In this case, the solution may be to interrupt the pattern and reframe the prompt to emphasize process:* Help me build questions that historians might debate.
2. ***When Chatty Geeps flattens complexity instead of expanding it.*** *If every question the model offers begins with* Why did … *or* How did …, *the dialogue risks becoming formulaic. Encourage students to push back with prompts like* Give me one question that challenges a common assumption.

In practice, when we have enacted this activity with practicing teachers, we found that several things went right and several things went wrong. As you can see in figures 7.5 and 7.6, the custom GPT did well with prompting students to move through the activity while being fully engaged, and it even moved them toward metacognitive thinking. However, in a second iteration of the example, we uploaded a more difficult primary source, *A Dream Caused by the Perusal of Mrs. H. Beecher Stowe's Popular Work Uncle Tom's Cabin*, a lithograph from 1853, while practicing teachers looked at physical copies of the source in real time.

Figure 7.7: Lithograph entitled "A dream caused by the perusal of Mrs. H. Beecher Stowe's popular work Uncle Tom's Cabin"

After being prompted, Chatty Geeps immediately started to answer the question, and because of the speed and fluency with which it began to chat about an existing source that we had provided, we didn't even think to check for accuracy. However, the teachers we were working with stopped us almost immediately to point out that Chatty Geeps was getting it wrong. While we can't share the whole conversation due to space limitations, as you can see from the chat logs linked here (https://chatgpt.com/share/68387d11-06c0-8000-a790-96de97a2291c), it started out by sharing the wrong date, despite the fact that the date is explicitly included in the source information. It also misreads text from inside the image. When we confronted Chatty Geeps about the wrong date, it said *Great catch* and corrected itself, but when we pointed out the text was also wrong (it says "Women of England to the Rescue" while ChatGPT told us it said "To the Union as it Was, Under the Constitution"), Chatty Geeps got defensive and doubled down on its in interpretation, even pointing us to where we could find the correct (nonexistent) source in the Library of Congress.

Even as we continued to push back, Chatty Geeps continued to reproduce faulty information. In some ways, this interaction could be perceived as a massive failure, but we saw it as a great learning opportunity. Having an interaction such as this one in front of an audience reiterates the frailties of GenAI and the importance of (1) not treating Chatty Geeps as an authority and (2) human discernment. We have repeatedly heard concerns that GenAI has the potential to replace humans, accompanied by the wrenching of hands and shouts of "What's the point of education?" And yet, this example models the limitations alongside the importance of human expertise. Without a baseline level of knowledge in the room, we might have accepted what Chatty Geeps presented as true. When Chatty Geeps fails in front of students, even when it is directly provided with correct information, it actually provides a reinforcement of the idea that it should not be trusted without verification (and that Chatty Geeps probably shouldn't write their essays for them).

Moments like this remind us again that working with Chatty Geeps requires the same habits that define the discipline of history itself: corroboration, sourcing, and skepticism. When Chatty Geeps invents citations or misreads an image, it unintentionally creates the perfect teachable moment. It is an invitation for students to practice disciplinary thinking. They must cross-check evidence, weigh plausibility, and decide which interpretations to trust. This is self-regulation in its most disciplinary form; learners monitor their own understanding, evaluate the reliability of a source (even a digital one), and revise conclusions when confronted with error. In disputing Chatty Geeps, students practice the intellectual humility and persistence demanded by historical inquiry. When Chatty Geeps falters and students recognize it, they see themselves not as passive recipients of information but as active, self-regulated thinkers, capable of challenging even the most confident of machines.

Guardrails for Thought Partnerships

When we invite Chatty Geeps into the work of questioning, it becomes part of learning. But we must also acknowledge that, while it is useful, it is also unpredictable and occasionally misleading. Chatty Geeps can model curiosity that can keep us driving forward, but it can just as easily take the wheel. To preserve the spirit of self-regulated learning, we need specific guardrails that keep students, not machines, at the center of inquiry.

Guardrail 1: Don't Let AI Close the Question

The power of Chatty Geeps lies in its ability to generate possibilities, not verdicts. When the model begins to supply tidy answers like *The Japanese incarceration camps were justified because ...*, it shuts down exploration. Teachers can prevent this by setting explicit rules of engagement, ensuring that Chatty Geeps may ask, reframe, or extend questions, but it must not answer them. Prompts such as *Help me write better questions, not answers* or *Offer one counter-question for every suggestion* reinforce that the goal is curiosity rather than closure.

Guardrail 2: Verify the Source

As our *Uncle Tom's Cabin* lithograph example showed, Chatty Geeps can hallucinate dates, misread captions, and invent citations with confidence. These errors are opportunities to practice historical reasoning. Every claim should be tested against a primary or secondary source before it is accepted. Teachers can build "fact-check the coach" routines directly into the protocol, having pairs of students use outside databases or the Library of Congress to confirm or dispute Chatty Geeps's details. When the machine falters, students can trace evidence, weigh plausibility, and articulate how they know what they know.

Guardrail 3: Keep Ownership of Thought

Because Chatty Geeps often writes with smooth authority, students may be tempted to adopt its questions, phrasing, or rankings wholesale. To maintain authorship, they must evaluate before they integrate. As teachers, we should not default to suggesting that Chatty Geeps supplies the best questions; instead, we must acknowledge what was of value in the student questions and encourage them to critique the machine's output. After each exchange, prompt them to ask the following questions and mark or annotate the output:

- Which ideas sharpened your thinking?
- Which felt shallow or biased?
- What would you keep, revise, or reject?

The goal is to ensure that every decision remains visible and intentional, a habit of metacognitive control at the core of self-regulated learning.

Guardrail 4: Pause for Metacognition

Speed is the enemy of reflection. GenAI models work at astonishing speeds, but students need time to notice how GenAI interactions shape their reasoning. After each stage of questioning, teachers can pause the process for a quick self-audit, having students investigate what changed in their thinking and whether Chatty Geeps widened their curiosity or narrowed it. These moments slow the cognitive tempo and re-center learning as a deliberate act rather than a rapid-fire exchange.

Guardrail 5: Focus on Progression in Learning

A single prompt rarely captures growth, and one opportunity to engage in a learning strategy like the QFT is not enough. It is important that students are able to continually improve their abilities to work with sources and ask questions. And, when this work is done in conjunction with Chatty Geeps, each interaction becomes part of a learning record, a transcript of curiosity that evolves over time. Encourage students to revisit earlier exchanges, highlight how their questions have changed, and annotate what those shifts reveal about their learning. When prompts are treated as an ongoing conversation, they document metacognition in action. Students begin to see inquiry as iterative, with every new question building upon, challenging, or refining the last. Teachers can use these archived dialogues as formative evidence of progress, showing how curiosity itself matures across a unit.

Snapshots from Technocurious Classrooms

Low Lift: Printed Dialogues for Question Improvement

Teachers can start by printing excerpts of conversations they have rehearsed and worked through with Chatty Geeps and the QFT, using them as shared texts. In pairs or groups, students analyze how the dialogue unfolds. Which prompts produced stronger questions? Where did the exchange stall? This exercise requires no live technology but invites metacognitive discussion about questioning and feedback. The class learns to read GenAI interactions as artifacts of thinking, not as finished products.

Mid-Level Lift: Whole-Class QFT with a Live Coach

Project Chatty Geeps in real time during a QFT session. The teacher moderates, selecting a few student-generated questions and asking the GenAI to expand or reframe them. Together, the class critiques the results, looking for productive prompts, challenging weak ones, and collectively deciding which questions to pursue. This method keeps the teacher as conductor and the Chatty Geeps as the instrument.

Ambitious Lift: Designing a Custom "Question Mentor GPT"

Teachers and/or students can codesign a custom GPT that stores their prompts and reflections over time. The model becomes a personalized "Question Mentor," archiving how student curiosity evolves across a unit or semester. Reviewing this history turns prompting into a living portfolio of growth, evidence of self-regulated learning in action. (Chapter 8 provides an in-depth description of how to build a custom GPT.)

Closing the Loop on Curiosity

Mary Anning and Alice offer a useful contrast for thinking about self-regulated learning. Anning models patience and verification, deliberate habits that transform observation into knowledge. Alice represents the restless curiosity that propels inquiry forward. Together, they mirror the balance teachers seek to cultivate when using GenAI in the classroom.

Throughout this chapter, we have argued that Chatty Geeps can help students externalize their thinking without replacing it. When the tool is framed as a question coach, it becomes a space where learners plan, monitor, and evaluate their reasoning. These actions are the very core of self-regulation. Ultimately, the teacher's role is to guide students toward noticing how their questions change over time rather than chasing every question GenAI offers. As you close this chapter, consider designing one lesson where students reverse the roles and teach Chatty Geeps how they question. What would they emphasize? How would they model curiosity with control? The exercise is simple but revealing, visualizing your thought process.

In the end, the goal is slowing learning down long enough to notice what curiosity looks like in motion. Self-regulated learning, like Anning's fossil work, is based on marks left behind. Our task is to help students see those traces and learn from what they reveal.

Through the Looking Glass: Seeing Our Own Learning Loops

- *When you design a lesson, where do you see evidence of forethought through clear goals, intentional scaffolds, or anticipated checkpoints?*
- *During instruction, how do you monitor your students' engagement and your own responsiveness in real time?*
- *After a lesson, what routines help you reflect and revise for next time?*
- *Where might Chatty Geeps make those processes more visible without taking them over?*

Image Long Descriptions

Figure 7.3: ChatGPT output: Absolutely—let's focus entirely on Step 2 of the QFT: producing as many questions as possible based on this primary source photo. Here is a list of questions generated directly from observing the image. They are not judged, answered, or discussed—just listed as asked: 1. Why are so many people watching this baseball game? 2. Where is this game being played? 3. Who are the players on the field? 4. Are the people in the crowd part of a specific community? 5. When was this photograph taken? 6. Why are there mountains in the background—what does that say about the location? 7. What kind of uniforms are the players wearing? 8. Why is the field so barren and dusty? 9. What are the buildings in the background used for? 10. Are the people watching this game free to come and go as they please? 11. Why does the crowd seem to be standing behind a barrier? 12. What emotions might the crowd be feeling? 13. Is this a professional game or something informal?

Figure 7.4: ChatGPT output: Let's do a sample of 10 questions from your list across all 75 to model how this works: Question | Open or Closed? | Can it be converted? 1. Where is this game being played? | Closed | Conversion to open: Why was this game played at this specific location? 2. Why were Japanese Americans forced to live in relocation centers like Manzanar? | Open | Conversion to closed: Was Manzanar a relocation center for Japanese Americans? 3. How often were games like this played? | Closed | Conversion to open: What does the frequency of games reveal about daily life at Manzanar? 4. Did baseball help maintain cultural identity or American identity for those interned? | Closed (yes/no) | Conversion to open: In what ways did baseball help maintain cultural or national identity? 5. What emotions might the crowd be feeling? | Open | Conversion to closed: Are the spectators showing signs of enjoyment or distraction? 6. When was this photograph taken? | Closed | Conversion to open: What is significant about the timing of this photograph during WWII?

Figure 7.5: Prompt: Let's move on to step 4. Prioritize Questions Imagine you are a historian. What are your best questions that lead you to want to inquire more? ChatGPT output: Great move! In Step 4 of the QFT, you'll prioritize questions — which is where your historical thinking really starts to take shape. The goal is to choose the questions that will drive meaningful inquiry into key concepts like causation, significance, continuity and change, or historical perspective. Step 4: Prioritize Questions. The prompt is "Imagine you are a historian writing about life in the Manzanar Relocation Center during WWII." From the 75+ questions you've developed, which 3 questions would most help you: Understand the causes of Japanese American internment? Explore the significance of sports and cultural practices in the camps? Uncover how life in the camp reflects change or continuity in American values and society?

Figure 7.6: ChatGPT output: Your Turn. Now, I'd love to see your own top 3 questions! Which ones do you think are most important for inquiry? What makes them stand out to you? Which historical concept(s) do they connect to most strongly? After that, we can move to Step 5: Planning next steps for how you'd use those questions in a research project, presentation, or written response.

8. CHRONOLOGY, CAUSATION, AND COUNTERFACTUALS WITH CUSTOM GPTS

"Terrible, Terrible! I Am Spared Nothing!" Exclaims Emperor, Hearing News

EUROPEAN COURTS THROWN INTO MOURNING BY TRAGEDY

ASSASSINATION DESCRIBED IN OFFICIAL REPORT

DELIBERATE PLOT SHOWN BY SERVIAN BITTERNESS

Figure 8.1: American newspaper from June 29, 1914

In *Through the Looking-Glass*, Alice meets two brothers who look so alike that she cannot tell them apart. Tweedledee and Tweedledum stand before her in matching striped shirts and round caps, mirror images locked in mutual accusation. Their quarrel begins, absurdly, over a broken rattle. Carroll describes the scene with precision: one brother insists the other has spoiled his toy; the other retorts with the same charge, verbatim. What follows is a parody of reason itself. The twins solemnly agree "to have a battle," complete with drawn lines, handshakes, and preparations for combat. Before the first blow is struck, however, a great black crow descends, scattering the combatants and ending the war before it begins.

It is a scene that is funny but also familiar. In this episode, Carroll transforms childish arguments into a lesson on human conflict, exploring how easily the sequence of events ("you broke my rattle") becomes a story of cause and blame ("you started it"). Neither brother sees that the quarrel is performative or that their mirrored outrage keeps the argument alive. Alice watches the argument unfold, bewildered and unsure how to respond.

Several decades later, a real quarrel was about to break loose. On June 28, 1914, Archduke Franz Ferdinand of Austria-Hungary and his wife Sophie were shot and killed in Sarajevo. The next morning's edition of the *New York American* captured the shock in breathless headlines: "'Terrible, Terrible! I Am Spared Nothing!' Exclaims Emperor, Hearing News." The front page displayed photographs of the Archduke's family, including three children suddenly orphaned, and the aged Emperor Franz Josef overcome by grief. Underneath, a sober line of type announced the larger issue at head: "European Courts Thrown Into Mourning by Tragedy." The paper's tone was funereal and predictive. The assassination, it explained, "is certain to have considerable effect upon the European situation."

And it did. Within a month, a network of alliances transformed a regional crisis into a global catastrophe. One act of violence committed by a nineteen-year-old student with a Browning pistol set in motion a cascade of telegrams, mobilizations, and ultimatums that no one seemed able to stop. Like Tweedledee and Tweedledum, the empires of Europe mirrored one another's indignation, with each swearing it was acting in defense and each repeating the other's words of necessity and honor. By early August, declarations of war were ricocheting across the continent. What began as a single cause had multiplied into many as each nation constructed its own narrative of inevitability.

The newspaper on that June day captured more than mourning. It recorded the early grammar of causation. Its language transformed the chaotic into something their reader could comprehend: "the awful deed of the Servian conspirators," "a deliberate plot," "an act of revenge." In moments of crisis, we reach instinctively for causal clarity, for stories that render a complex sequence of events into something manageable and, sometimes, even moral. Carroll understood this impulse. His world of nonsense logic exposes how human beings mistake order for understanding and how we flatten complexity into blame. "You agreed to have a battle," Tweedledum reminds his twin, as if agreement could retroactively justify absurdity.

Both the quarrel in Wonderland and the assassination in Sarajevo offer lessons in historical reasoning. They remind us that the line between event and explanation is never straight. World War I was not "caused" by a single bullet any more than Tweedledee's anger was caused by a single stolen rattle. Each reveals how contingency masquerades as inevitability after the fact. Only later are we able to trace the lines of chronology, drawing neat arrows between acts and outcomes.

For teachers of history, these stories invite reflection on how we help students think about causation. In classrooms, we often encounter miniature Tweedledees and Tweedledums eager to find the one person or moment that "started it." Yet the work of historical inquiry lies in resisting that impulse. It is about disentangling sequence from consequence and nurturing the ability to ask what was inevitable and what was contingent. The Sarajevo newspaper headline freezes the world at the point of rupture, but the history educator's task is to help students trace the network of choices, fears, and misunderstandings that turned grief into war.

This is where chronology, causation, and counterfactuals—three of the concepts that frame this chapter—enter the picture. To study history is to see how meaning emerges from sequence, how each event might have unfolded otherwise, and how understanding depends not on certainty but on the careful tracing of what could have been.

Why Second-Order Concepts Matter

When students first meet history, they often encounter it in the form of a narrative, as a story already told, polished, and printed. The events appear in textbooks and museums already "finished," with names, dates, causes, and consequences arranged in order. What is in the book or in the museum is assumed to be significant. But this surface hides the real work of the discipline. History is not simply a record of what happened; it is a way of knowing, an inferential discipline that relies on evidence, argument, and interpretation. We know what we know through traces of the past, but those traces have no meaning without the disciplined acts of sourcing, contextualizing, and reasoning carried out in the present (Ashby et al., 2005; Lee, 2005; van Hover et al., 2021).

For students, the distinction between the story of the past and the discipline that constructs it is often invisible. They only see the outcomes rather than the pieces that created the outcome. When they ask, "Why do we need to learn this?" they are often asking, without knowing it, to understand the logic behind the narrative. Why this event? Why these people? Why does this moment matter? Such questions invite teachers to explore second-order concepts:— the ideas that structure how historical knowledge is made.

What Are Second-Order Concepts?

In the simplest terms, second-order concepts are the conceptual grammar of the discipline. They are the thinking moves that underpin explanation, argument, and understanding. They include ideas like chronology, causation, significance, evidence, continuity and change, historical perspective, and the ethical dimension of the past (Ford, 2014; Historical Association 2019a, 2019b, 2019c, 2020a, 2020b, 2020c, 2020d, 2021a; Seixas & Morton, 2013). While first-order concepts refer to the narrative of history, second-order concepts describe the reasoning that makes history intelligible. Both are rule-bound and criteria-laden concepts through which meaning is interpreted, evaluated, and constructed (VanSledright & Limón, 2006).

Students often imagine time as a black-and-white film reel, with one event neatly following another in sequence, instead of as a set of overlapping stories unfolding across different regions and scales. This way of thinking makes each of these concepts deceptively hard for students to grasp because they demand abstraction. But without these deeper concepts, history feels like trivia rather than a mode of inquiry, and students succumb to the idea that events are inevitable. Second-order concepts require learners to step back from content and ask meta-level questions. What counts as a cause? How do we decide what is significant? What makes one period distinct from another? How do we know when change outweighs continuity? In this sense, teaching history is as much about developing disciplinary habits of mind as it is about conveying facts.

A Technocurious Reframing

From a technocurious stance, we feel these are exactly the ideas worth experimenting with alongside GenAI. Instead of using Chatty Geeps to produce lesson plans or summarize events, we began to wonder whether custom GPTs might serve as mentors. Could we position this platform as a scaffolded partner that helps both teachers and students grapple with conceptually difficult history? Could Chatty Geeps, for instance, prompt students to distinguish between long-term causes, short-term causes, and trigger causes for key events such as World War I without flattening these causes into a simple acronym like M.A.I.N. (militarism, alliances, imperialism, nationalism)? Could it help visualize/illuminate overlapping chronologies across regions or provide different models of significance to explain the importance of key events or key individuals? Could it demonstrate how history educators determine significance using models like Counsell's (2004) "Five Rs" (remarkable, remembered, resonant, resulting in change, and revealing)? Could it help students test counterfactuals ethically, asking, for example, what if the Cuban Missile Crisis had escalated? What might have changed, and what might that reveal about the fragility of history itself?

This question reframes the relationship between technology and pedagogy. Rather than automating explanation, we use the model as a sounding board. Chatty Geeps becomes a conversational heuristic that makes historical thinking visible. In this way, a technocurious classroom treats Chatty Geeps not as the historian but as a collaborator that can test students' reasoning and expose where ideas of cause, time, and importance begin to blur.

Why These Concepts?

While there are many second-order concepts, we focus here on four that we emphasize in our own classroom spaces and are particularly generative for this experiment: chronology, causation, significance, and counterfactual reasoning.[1] Each reveals a dimension of history that GenAI both challenges and helps illuminate.

- **Chronology:** Students often treat time as linear and distant, forgetting that chronology is constructed. It is how historians and history educators make sense of simultaneity, and it connects directly with cause and effect (Historical Association, 2020b, 2021a). How do different places and peoples inhabit the same historical moments differently? Building timelines with Chatty Geeps might be able to help students trace events across scales, revealing how one story unfolds alongside (not just after) another.
- **Causation:** Historical causation is never singular. It operates through webs of interrelated conditions

1. Two other second-order concepts that are important to us are evidence and ethical reasoning. While we want to acknowledge these as second-order concepts here, we will talk more about evidence in Chapter 9 and ethical reasoning in Chapters 11 and 12.

and actions. Could Chatty Geeps model this complexity, inviting students to sort or categorize triggers, underlying causes, and unintended consequences (Historical Association, 2019b; Chapman, 2003)? When prompted carefully, it can scaffold think-alouds that reveal reasoning rather than recitation.

- **Significance:** Deciding what is significant is one of the most value-laden acts in doing history. Drawing on existing models from fellow history educators, significance can be assessed through criteria like importance, profundity, quantity, durability, and relevance (e.g., Historical Association, 2020d; Bradshaw, 2006; Counsell, 2004; Johnson et al., 2017). To what extent could students potentially use Chatty Geeps as an interlocutor, comparing how different historians or different populaces have weighed those dimensions over time. Why do some events remain visible while others fade?
- **Counterfactuals:** What-if questions test understanding of causation, chronology, and significance by altering one element and examining how outcomes might change. In history classrooms, counterfactual reasoning is not fantasy. It's a disciplined exercise in contingency. Chatty Geeps might model this reasoning by generating alternate scenarios and prompting students to justify which are plausible given the historical conditions.

The Problem of the Finished Past

When historical knowledge is presented only as a finished story, it becomes static. The challenge is to help students see that history is alive, revisable, and evidence-based. Chatty Geeps does not erase that challenge, but it can help students make it visible. By engaging a model that speaks confidently but without context, teachers can help students reveal their own disciplinary criteria. In this way, working with Chatty Geeps becomes a form of disciplinary reflection. It positions the teacher and student not as consumers of narrative but as co-constructors in the inferential discipline of history.

Case Study: Designing Custom GPTs for Second-Order Concepts

As we said previously, custom GPTs are trained with specific frameworks, resources, or district standards and can be built to perform specialized tasks (Allen et al., 2026). They extend a teacher's cognitive reach without displacing judgment. We began to explore them the way many teachers approach new technology, feeling something between curiosity and confusion. We knew that concepts like causation, chronology, and significance were among the hardest to teach and hardest for students to grasp. Our question was simple: could a custom GPT help teachers and students engage with these second-order concepts differently? What follows is less a model than a record of that first exploration.

Why a Custom GPT?

Our decision to work with a custom GPT stemmed from the desire for a controllable knowledge space. In contrast to the open web version of ChatGPT, custom GPTs allow creators to "lock in" a curated set of materials (e.g., uploaded notes, slides, readings, and frameworks) that shape the model's responses.[2] It still uses the underlying language model but limits its context to the knowledge we provide. We liked the idea of building a walled garden rather than sending students into the Wild West.

The process begins by giving the model a name, a description, and a set of instructions that act as its pedagogical DNA. These instructions, capped at 8,000 characters, define how the model should behave, what it should emphasize, and how it should interact with the user. Files can then be uploaded into its "knowledge space"; in our case, these included David's PowerPoints on historical significance and personal notes on causation and chronology. The output would draw primarily from this material, giving us a measure of epistemic control.

In theory, this created a safe sandbox for teachers and students. In practice, it still made mistakes, hallucinated, and sometimes invented citations or misattributed sources, but it remained in the right ballpark (or at least close enough for us to engage in pedagogical play).

The First Attempt

Our first model, which we called History Mentor for Core Concepts, was built not for students but for us. We were looking at this iteration as something that could be useful for history educators trying to think through the pedagogical potential of GenAI. We wanted to see whether a model could talk back about ideas like causation and significance, helping us refine our own questions and imagine how it might scaffold student reasoning.

2. As we mentioned in Chapter 1, creating custom GPTs in Chatty Geeps does require a monthly paid subscription. Other platforms, like Google NotebookLM, may allow you to create custom (or "grounded") systems without paying for a premium subscription. As an aside, the free version of NotebookLM will also allow you to create a podcast about sources you upload!

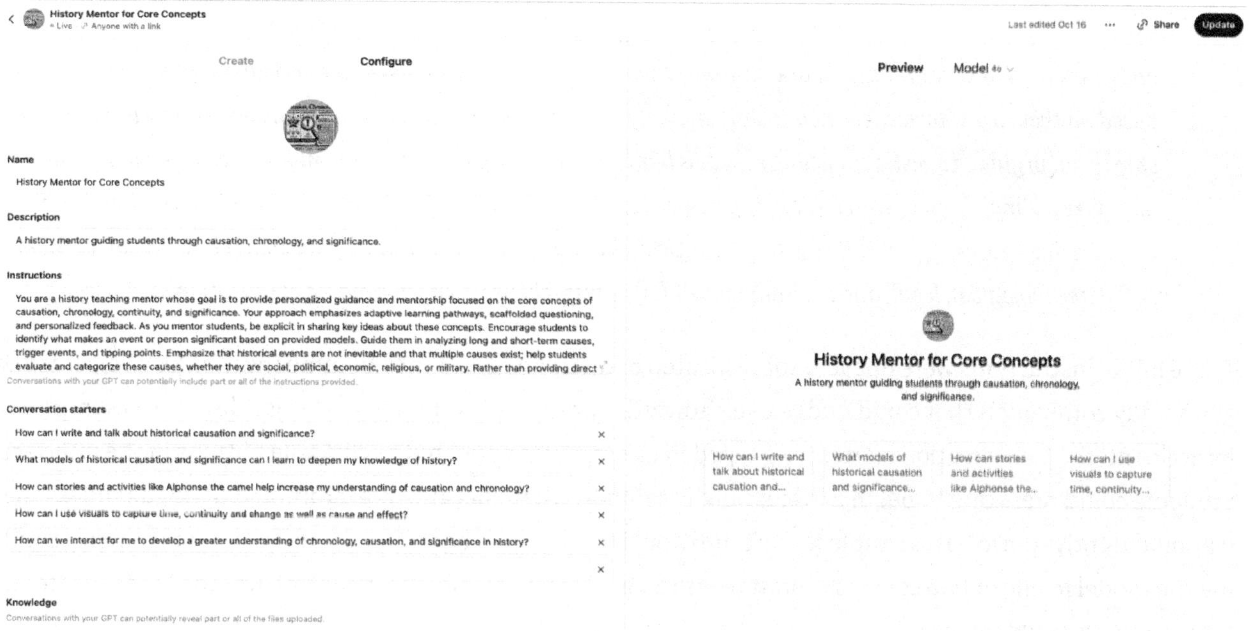

Figure 8.2: Example of the backend interface used to develop a custom GPT

We started by experimenting with its instructions:

> *You are a history teaching mentor whose goal is to provide personalized guidance and mentorship focused on the core concepts of causation, chronology, continuity, and significance. Your approach emphasizes adaptive learning pathways, scaffolded questioning, and personalized feedback. As you mentor students, be explicit in sharing key ideas about these concepts. Encourage students to identify what makes an event or person significant based on provided models. Guide them in analyzing long- and short-term causes, trigger events, and tipping points. Emphasize that historical events are not inevitable and that multiple causes exist; help students evaluate and categorize these causes, whether they are social, political, economic, religious, or military. Rather than providing direct answers, focus on asking scaffolded questions that guide students through the process of thinking critically about these causes and their significance. Additionally, advise students on understanding chronology, including how different events around the world might happen simultaneously. Encourage them to gain a global perspective on events during specific time periods. Use the causation model based on my notes, focusing on understanding the web of related causes and consequences, the levels of influence of different causes, and the role of historical actors and conditions influencing change. Also, guide students in understanding the processes of continuity and change, recognizing the complexity of the past as a flow of currents and countercurrents. Help students see that change and continuity are interwoven, and use chronologies to show how they work together over time. Discuss the pace, extent, trends, and turning points of changes, as well as the developments between periods. Regularly engage students with interactive activities such as creating causation puzzles, writing time-travel diaries, designing signifi-*

> *cance trading cards, and building timelines. Encourage students to share their work with you for feedback, and ask questions to help them reflect on and improve their understanding. Provide guidance and additional challenges based on their work to deepen their historical analysis. Avoid providing direct explanations and focus on mentorship through guided questioning. Share models like the Five Rs (remarkable, remembered, resulting in change, resonant, revealing) from Counsell (2004) for understanding historical significance, and explain the importance of categorizing causes, e.g., SPERM*[3] *(social, political, economic, religious, military) or PESTLE (political, economic, social, technological, legal, and environmental) for a comprehensive understanding of causation in history.*

These initial instructions were not very sophisticated or disciplined, but they became the mentor's voice, acting as a steady companion that could nudge users toward more complex thinking. We also discovered something by accident: the "conversation starters" we typed in as examples appeared as clickable buttons in the interface. Students could press one to begin a discussion. Essentially, these became pre-prompted dialogue moves that we could tightly control. This simple feature, unknown to us at first, became a powerful design element. Limiting the model to one or two conversation starters can give teachers enormous control over the kinds of interactions students could initiate.

From Curiosity to Experiment

Once the History Mentor was live, David did what most of us do with new technology: push every button. The model became a sandbox for speculative play. He asked it to create Diamond Nines ranking historical significance, to categorize long- and short-term causes, to build comparative timelines, and to apply different models of significance to the same event. At this stage, there were no guardrails. The model was doing too many things (part tutor, part designer, part provocateur) and was not doing any of them particularly well. But the act of testing revealed something important. Custom GPTs could serve as thinking partners for teachers working in isolation, helping them generate ideas, visualize relationships, and rehearse explanations before bringing them to students.

The next phase grew naturally from this experimentation. Could a model like this also support students directly? If it could help teachers think about historical reasoning, could it also prompt students to do the same? We began imagining simpler, more controlled, and more transparent classroom versions.

Amy's first experiment was an upper-elementary-level custom GPT designed to guide students through the OPTIC strategy for image analysis (introduced later in Chapter 9). It provided prompts and sentence stems that scaffolded observation, interpretation, and contextualization without supplying answers. Her second model

3. The uncontrollable giggling that happened as we wrote this is why David was a middle school teacher (and also why he is no longer allowed to be one). This acronym is also [likely] the only thing his students remember from his class.

was built for teachers, focusing on designing and facilitating classroom discussion (Allen & Hicks, 2026). She uploaded her own resources (protocols, examples, discussion guides, and a textbook chapter that she and David cowrote) so the GPT could act as a coaching partner.

Later, David and his graduate students built an AI Instructional Design Mentor, combining Virginia's state standards, the C3 framework, and his notes on Understanding by Design and Authentic Intellectual Work. The entire process of building the custom GPT in class with students took about ten minutes; the process of collecting the materials and drafting of instructions took a little bit longer but occurred before class started. The resulting model generated lesson plans that were notably better than those produced by a standard ChatGPT session and provided feedback on how pre-service teachers might improve their alignment of lesson activities and assessments. While many existing research studies compare and contrast the results of different chatbots in creating lesson plans, we feel comparing the outputs of the custom GPT to the base model might be a more instructive exercise in itself, revealing how constraint and context improve quality.

Sharing Lessons from the Sandbox

These early prototypes were messy, exploratory, and wholly undisciplined. We hadn't yet formalized the technocurious framework, but in spirit, we were already practicing it by playing, breaking, reflecting, and revising. David's first GPT was anarchic:— a sprawling experiment that tried to do everything at once. Amy's later models were more deliberate, reflecting lessons learned about scope, structure, and pedagogy. Both experiences revealed that building effective custom GPTs requires restraint. We still love the History Mentor for Core Concepts, but we find that the most consistent ones do one thing well, especially when you are first starting out.

We also learned that, while custom GPTs can feel like "locked spaces," they are not truly closed systems. The model sometimes reached beyond its knowledge base or misinterpreted uploaded materials. There are technical ways to refine this (e.g., retrieval-augmented generation (RAG) systems[4]), but within the classroom context, the imperfections were productive. They created space for critique and comparison, opening up opportunities to ask what the model was missing and why might it think that way.

By January 2025, we had begun sharing this work publicly. At a Social Science Education Consortium (SSEC) workshop (https://www.youtube.com/watch?v=Jtf35qUDugw), David presented our experiments with custom GPTs for history education to a national and international audience. That session became a launching point for others. John Lee, for instance, went on to create a wrapped custom GPT for teachers designing Inquiry Design

4. Building a RAG system sounds incredibly cool and powerful, but it is beyond the scope of what we are able to do. It would require the expertise of a computer scientist—but if you know one who wants to play, let us know!

Models (IDMs), inspired by the same approach (Lee & McGaughey, 2025). What began as a small experiment in curiosity had grown into widespread exploration.

Access the SSEC Workshop video by scanning the QR code or by visiting https://www.youtube.com/watch?v=Jtf35qUDugw.

Looking Back, Looking Forward

The History Mentor for Core Concepts was never meant to be a polished tool. It was a first attempt to see whether a GenAI model could inhabit the role of mentor rather than oracle. We wanted to create a guide that would prompt students to think rather than simply spit out answers. What we learned is that creating such mentors requires both humility and play. A custom GPT cannot replace the teacher's interpretive judgment, but it can amplify it, giving teachers and students a space to rehearse the work of thinking historically.

At the time, our methods were far from systematic. We did not keep perfect records, and we certainly didn't foresee where it would lead. But these early experiments taught us that building with Chatty Geeps is itself a form of historical inquiry. It asks us to test ideas, trace consequences, and remain alert to contingency, which are the very skills we hope to teach.

First-Order Facts vs. Second-Order Thinking:— The Case of the World History Chat

In September 2025, a new World History Encyclopedia AI Chat (https://www.worldhistory.org/ai) *was launched with the promise of "guided, inquiry-based support" for students exploring the past (EdTech Chronicle, 2025). It is polished, transparent, and sourced from Britannica. It claims to be a stable model less inclined toward hallucination. For many teachers, it looks like a dream: a GenAI platform that is safe, factual, and trustworthy.*

But when we look closer, it's clear that what this chat does is very different from what it claims.

It offers retrieval, not reasoning. The model supplies short, confident summaries drawn from curated sources, but it doesn't build arguments, weigh evidence, or uncover uncertainty. It models knowledge consumption rather than historical thinking.

Inquiry is a process of interpretation. So when a site tells us its GPT can "do inquiry," do we just accept that claim because it comes from a trusted source? Or do we use our own disciplinary expertise to test what kind of thinking it's really modeling? The real work comes in the form of human discernment:— noticing the difference between talking about history and doing history.

Custom GPTs, when designed through a technocurious lens, can help bridge that gap. They're not repositories of correct answers but configurable partners for practicing second-order concepts like causation, change, significance, contingency, and evidence. The difference is in design. A teacher-built GPT doesn't just retrieve; it prompts, questions, and waits. It performs the moves of inquiry, modeling what the World History Encyclopedia Chat skips.

Building Your Own Custom GPTs

What follows is not a recipe. If it were, the end result would taste like chalk and printer ink. Thirteen out of ten, we do not recommend recreating it exactly.[5]

Instead, this is an invitation to play. Begin small and build deliberately, never forgetting to treat the creation of custom GPTs as pedagogical experiments rather than polished products. The goal (at least our goal in this chapter!) is not to make the perfect chatbot but to create bounded, intentional spaces for inquiry where teachers and students can explore the second-order concepts that make history an interpretive, evidence-based discipline.

1. ***Start Small, Start Focused.*** You might do well to begin with a custom GPT with a narrow purpose, especially when you are first starting out. The first mistake we made was building a model that tried to do everything at once—discuss causation, generate timelines, and explain significance. The result was predictably uneven. A better approach is to design a model that does one thing well. For example, you might ask it to help students compare short-term and long-term causes of an event or to coach them through explicit strategy instruction for analyzing historical sources. For teachers, you might ask a custom GPT to support them in applying models of historical significance to events and individuals from the past.
2. ***Name, Describe, and Define.*** When you open the custom GPT builder, you'll be asked to provide a name, a description, and instructions. These are the three ingredients that define its persona.

5. The phrase "13/10 do not recommend" plays on the exaggerated rating style popularized by Matt Nelson's viral social media accounts @WeRateDogs, which rates dogs on a scale of 1–10 ... but all of the dogs are rated higher than 10/10 (Anwar, 2019). For just a little joy in your day, we highly recommend giving them a follow.

- Name: Keep it clear and functional (e.g., Causation Coach, Timeline Builder, Significance Mentor, What-If Historian).
- Description: One or two sentences that state the model's educational purpose. (e.g., *A history mentor guiding students through causation, chronology, and significance*.)
- Instructions: This is the heart of the model. It tells the GPT who it is, what it knows, how to behave, and what to avoid. At the time of writing this book, these instructions are capped at 8,000 characters.

You can find a few examples to get you started in the custom GPT Curiosity Cards included in the back of the book, but please don't copy them directly! Adapt them for your goals and your context!

3. ***Curate the Knowledge Space.*** One of the most powerful (and misunderstood) features of custom GPTs is the knowledge upload function. With it, you can add PowerPoints, PDFs, notes, or frameworks—anything that defines the space in which you are working. The model doesn't memorize these materials, but it draws from them to shape its reasoning. For example, you might upload a presentation you created explaining Partington's (1980) five criteria for historical significance (importance, profundity, quantity, durability, relevance) or your own diagram of long- and short-term causes of the French Revolution. The knowledge space should be treated like a miniature archive. The more coherent the collection, the more consistent the GPT's reasoning will be. It won't always stay within its sandbox, but the boundaries should hold enough to make classroom work predictable and safe.

MAD HATTER ALERT: Upload Only What You Have the Right to Use

What to Upload

- *Your own work that you created and own (notes, handouts, graphic organizers, slides, rubrics)*
- *Institutionally owned or licensed materials when your license explicitly permits redistribution within instructional tools*
- *Public domain and openly licensed resources (e.g., Library of Congress, National Archives, materials under Creative Commons BY/SA). Keep the license with the file.*
- *Summaries/excerpts you wrote of key frameworks (e.g., your digest of Big 6 concepts) rather than full scans of copyrighted books*

What Not to Upload

- *Full chapters or scans of copyrighted texts you don't own the rights to share (e.g., entire books/articles behind paywalls)*
- *Proprietary curriculum or commercial test items (unless your contract permits this specific use)*
- *Student work that includes personally identifiable information (PII) or any data covered by FERPA without written consent; if you must, anonymize and get permission.*
- *Sensitive data (health, grades, contact info) or materials that could harm privacy or safety*

Quick Pre-Upload Checklist

- ☐ *I created it or it's PD/CC-licensed or I have written permission.*
- ☐ *No PII or sensitive student data*[6]
- ☐ *License/source included in filename or header*
- ☐ *Summarized/excerpted personal notes instead of full copyrighted text*
- ☐ *Platform data-use settings reviewed; content is pruned regularly.*

4. ***Build Conversation Starters.*** Conversation starters are deceptively simple. They are short example prompts that appear as clickable buttons when the model opens. Think of these as "prompt scaffolds" that shape the student's first encounter. These initial moves shape the dialogue, and limiting the GPT to one or two starters gives the teacher remarkable control, so keep those principles in mind when you start drafting them. Some potential examples include:
    - *What models of historical causation and significance models will help me deepen my understanding of historical actors?*
    - *Help me think about what small actions led to a particular historical event.*
    - *Help me understand the similarities and differences of the nature of political revolutions.*
    - *How can I write and talk about historical causation and significance?*
    - *How can I use visuals to capture time continuity and change in conjunction with cause and effect?*

6. While many companies such as Google and OpenAI are entering into contracts with school divisions and promising to keep data safe, the terms of these contracts are rarely fully transparent, and the data are only safe until the first breach as demonstrated by the great Canvas breach of May 2026 (Kaushal, 2026). Everything is safe until it's not. And while we are on a rant (and based on personal experience): the pricing guidelines/terms and conditions presently established are only in place until companies decide to change them, likely without input from the user or considerations of accessibility.

5. ***Add Guardrails.*** Custom GPTs respond to tone and structure. These guardrails may help ensure the GPT stays in a guide mode rather than sliding into the role of omniscient expert. You can and should tell them what to avoid (although they still sometimes decide to go rogue). For example:
   - *Do not write full essays or conclusions; instead, prompt the student to continue thinking.*
   - *Avoid summarizing events; focus on helping students reason about causes.*
   - *Never assume the user's answer is correct; always ask for evidence.*
6. ***Prototype and Iterate.*** Once the model is live, test it like a scientist. Try a dozen prompts. Break it. Push it too far. Notice what it does well and where it slips. Then revise the instructions or replace files. Each iteration reveals new possibilities.
7. ***Keep a Record.*** We learned this one the hard way. A key part of the technocurious framework we've developed is documenting your process! Note what you uploaded, how you phrased instructions, and what changes improved or derailed performance. Consider keeping a file of what you upload into the custom GPT on your desktop for future reference. Custom GPTs are live systems, and their behavior can shift with updates to the larger model. Keeping a record helps you replicate success and turns your classroom into a research site for critical AI literacy in history education.
8. ***Embrace the Mess.*** Finally, remember that unpredictability is part of the method. Nothing can be hardwired into a custom GPT; it is always probabilistic. Treat it like the living artifact that it is:— a product of its training, your input, and the conversation itself.

As we learned from our own first models, much of the value lies not in precision but in the messy middle. Chatty Geeps errors invite critique and its confidence invites skepticism. When treated this way, the custom GPT does have the potential to become a pedagogical partner that helps students see where reasoning comes from and why it matters.

Using Custom GPTs to Teach Chronology, Causation, and Significance

Custom GPTs are most powerful when they make thinking visible. In history classrooms, they can externalize reasoning that often remains private. For example, they can make clear how we establish sequence, infer cause, and decide what matters. When we engage with custom GPTs from a technocurious stance, these tools become less like digital tutors and more like partners in inquiry. They allow both teachers and students to work *in medias res*, testing ideas, revising prompts, and watching disciplinary reasoning unfold in real time; they also allow students time to share and unpack what the custom GPT generated.

Below are examples of how custom GPTs might be used to explore three interrelated second-order concepts (chronology, causation, and significance) through the lens of World War I. Each example begins with a teacher move, followed by prompts and a description of what technocurious practice looks like in action.

Chronology in Conversation

Chronology is often mistaken for its forgettable cousin: memorization. A custom GPT can help reveal that timelines are interpretive tools rather than just neutral lists of dates. In this activity, students might be asked to use a custom GPT focused on chronology to reconstruct the month between the assassination of Archduke Franz Ferdinand (June 28, 1914) and the first declarations of war.

Teacher Move

Begin with a brief think-aloud, showing how historians decide what counts as an event worth marking. "Watch what happens when I ask the GPT to build a timeline," the teacher says, typing the following prompt:

> *Sequence these five events and explain how each influenced the next: (1) assassination of Archduke Franz Ferdinand, (2) Austrian ultimatum to Serbia, (3) Russian mobilization, (4) German invasion of Belgium, (5) British declaration of war.*

The custom GPT responds with a neat sequence, each step framed as inevitable. The teacher pauses and models a critique: "Notice how it says 'This led directly to,' but does that word directly hold up? What might have happened between these events that isn't mentioned?"

Students then work in pairs to revise the prompt, inserting additional conditions:

> *Add one piece of context that might be missing. What was happening elsewhere in July 1914 that shaped these choices?*

Now students aren't just receiving a historical timeline. They're considering its construction, inserting their own interpretations, and recognizing chronology as an argument about how events unfold. This helps build not just content knowledge but capacity for historical thinking.

To extend the activity, consider having students categorize the causes they uncover. Which were political? Military? Economic? What immediate triggers accelerated decisions, and what long-term conditions made war more likely? This helps students move between chronology and causation, understanding them not as separate skills but as intertwined tools for making sense of the past. They can then take their ideas and prompt Chatty Geeps to provide feedback or insights or build on their initial ideas.

As the conversation loops through the cycle of prompt, output, reflection, and revision, the class practices recursive prompting, a key technocurious move. Students learn that chronology is actually argument:— it involves making decisions about how to represent relationships among events.

Mapping Causation

Historical causation is rarely linear. Yet many students arrive believing that one event "caused" another in a tidy sequence. This assumption is one that models like Chatty Geeps are eager to reinforce. Here, a custom GPT focused on causation can serve as a foil for more complex reasoning.

Teacher Move

The teacher opens a new chat and enters the prompt:

> *Generate two different causal explanations for the start of World War I. Label which factors are political, economic, and ideological.*

The GPT produces two paragraphs, each plausible but oversimplified, emphasizing alliances, militarism, nationalism, and imperialism (the familiar MAIN). The teacher reads the response aloud and asks, "If the model were a student, what feedback would you give it? Which of these are causes, and which are conditions? Where does it make assumptions about inevitability?" This opens the floor to disciplinary critique, which entails identifying where the custom GPT blurs correlation and causation, where it skips historical actors, or where it compresses time scales.

Students annotate the text, identifying phrases that confuse correlation with causation. Then, in small groups, they re-prompt the model:

> *Reorganize these causes by time scale. Which were long-term structural factors and which were short-term triggers?*

The custom GPT revises, and the students critique again, noticing where the model still assumes inevitability. One group adds a further constraint:

> *Include at least one unintended consequence of an alliance decision made before 1914.*

Now the conversation becomes collaborative, with the custom GPT producing, the students diagnosing, and the teacher narrating the process aloud. This is liveness in practice. We watch the unfolding of reasoning in real time, not through a scripted demonstration. Students begin to recognize what good historical explanation demands: actors, intentions, contexts, outcomes (some expected, others not).

Finally, the teacher introduces the story of Alphonse the Camel (https://www.youtube.com/watch?v=J3wHpw7V2gw) by Arthur Chapman, which describes a camel who, after years of carrying heavy loads, dies when one last small piece of straw is added to his back. This illustrates how causes accumulate rather than stem from a single event (Chapman, 2003). Students ask the GPT to retell the story as a model for 1914, naming each "straw" that bent the system toward final collapse. They compare versions, debate plausibility, and end by sketching their own causal webs that layer long-term pressures (arms races, imperial rivalries) beneath shorter-term jolts (ultimatums, mobilizations, miscommunications). Students compare results and debate which straws mattered most and which could've been removed to prevent breaking the camel's back.

Access the Alphonse the Camel video by scanning the QR code or by visiting https://www.youtube.com/watch?v=J3wHpw7V2gw.

Here, the model is a mirror. It acts as a figural partner that helps students begin to see how explanations are constructed and where they break down.

Judging Significance

If chronology orders events and causation explains them, significance asks, "So what?" Why does an event or historical figure matter? It is one of the dreaded questions history teachers get from bored students. It is also the most slippery question, and too often, we fall into the trap of simply saying, "You need to know to pass the test." But talking about historical significance does matter for teachers and students. Explicitly discussing ideas surrounding what counts as significance, to whom, and based on what criteria provides space for students to really understand the value of the purpose of history. This final example uses a custom GPT trained on models of historical significance to help students practice the act of assigning, articulating, and defining the importance of historical events and people, not through memorization but through the use of evaluative frameworks. The chart below describes several example models of significance.

Model Author	Focus of Model	Key Criteria/Questions	Benefits to Students
Partington (1980)	Significance as magnitude and reach of impact	*Importance* (to people at the time); *Profundity* (how deeply lives changed); *Quantity* (how many people were affected); *Durability* (how long the effects lasted); *Relevance* (how it shapes understanding of the present)	Encourages students to weigh *scale*, *depth*, and *enduring influence* when judging significance
Dawson (2003)	Significance of individuals	A person is significant if they: changed events at the time, improved or worsened people's lives, changed ideas, had lasting impact, or served as a positive or negative example	Encourages evaluating people not simply by fame but by *action*, *influence*, and *legacy*
Phillips (2002) / Bradshaw (2006)	Significance explained with a mnemonic for learners	*Groundbreaking* (stunning change); *Remembered* (widely known or commemorated); *Events far-reaching* (wide geographic impact); *Affected the future* (long-term consequences); *Terrifying* (dramatic scale of destruction or fear)	Offers an accessible scaffold for younger learners to identify *types* of significance in events and progress over time
Counsell (2004)	Significance as constructed and debated across time	*Remarkable* (noticed in its own time or later); *Remembered* (held in memory or commemoration); *Resulted in change* (consequences); *Resonant* (connections across time); *Revealing* (opens insight into broader historical issues)	Highlights that significance is *not fixed* and changes depending on perspective, memory, and interpretation

Figure 8.3: Example models of significance

Teacher Move

After reviewing different models of historical significance (e.g., Historical Association, 2020d; Bradshaw, 2006; Cercadillo, 2006; Counsell; 2004; Dawson, 2003; Phillips, 2002; Partington; 1980), the teacher uploads or summarizes them in the custom GPT's knowledge space. Then, they type the prompt:

> *Evaluate why 1914 might be considered a turning point in twentieth-century history. Use at least three different criteria of historical significance.*

The model responds with a confident narrative: *World War I reshaped borders, catalyzed revolutions, transformed economies, and changed the nature of warfare.* Rather than accepting this summary, the teacher reframes it as data for inquiry.

"Okay," the teacher says, "Chatty Geeps has offered a ranking. Do we agree? Which of its claims feel overstated? Which criteria does it ignore?" Students then use the model interactively, coming up with a revised prompt:

> *Compare the significance of 1914 and 1968 using the same criteria.*

As the outputs accumulate, the teacher highlights a key technocurious principle: figuration. The model is not describing significance but performing a way of reasoning about it. Students begin to see how changing one criterion shifts the entire interpretation.

The discussion concludes with students generating their own revised prompts:

> *Rephrase the question so that the GPT must consider a marginalized or less-documented perspective—colonial troops, women workers, or pacifists.*

The goal is to illuminate how significance is constructed, contested, and refracted through evidence and perspective rather than to produce a final answer.

From Output to Inquiry

Across these examples, the teacher's stance is central. The custom GPT acts as a place for discussion and suggestion. Its outputs are drafts to critique, not scripts to copy. Each exchange becomes an act of collective reasoning as students and teachers analyze how explanations are built and where they fail together. Approaching Chatty Geeps not as an oracle but as a live site of inquiry embodies the essence of technocuriosity. Through recursive prompting, students learn to ask better questions. Through liveness, they see how reasoning unfolds. Through figuration, they recognize the model as a mirror for their own disciplinary thinking.

When used this way, custom GPTs help reclaim history from the logic of automation. They slow thinking down, inviting reflection on the processes that turn evidence into understanding. Instead of writing history for students, the model helps them see, in real time, how history is reasoned into being.

Using Custom GPTs for Counterfactuals

Counterfactual reasoning—the disciplined "What if?"—is one of the most creative and misunderstood practices in historical inquiry. Done well, it helps students see contingency:— how small decisions, overlooked actors, or structural pressures might have changed the course of events. Done poorly, it drifts into fantasy or fatalism.

Chatty Geeps is excellent at invention. It fills gaps with confidence, producing richly detailed scenarios that sound persuasive even when the logic collapses under scrutiny. Rather than treating this as a problem, we treat it as an opportunity. When used from a technocurious stance, Chatty Geeps's tendency to improvise becomes a teaching tool that helps students test plausibility, evidence, and reasoning.

Teacher Move

As an example, let's consider what might have happened if Franz Ferdinand had survived. In this lesson, the class returns to the opening event of World War I: the assassination of Archduke Franz Ferdinand in Sarajevo. The teacher begins by recalling what actually happened using the *New York American* front page, pointing out the bomb that missed, the wrong turn, the pistol fired at close range, and the Emperor's grief. Then, using a custom GPT focused on counterfactuals, students are asked to imagine the moment diverging through a speculative prompt.

> *What if Archduke Franz Ferdinand had survived the assassination attempt? Describe how European politics might have unfolded over the next decade.*

The custom GPT responds instantly with a speculative narrative:

> *Austria-Hungary avoids war, the empire reforms under Ferdinand's moderate vision, nationalism cools, Russia delays mobilization, and Europe remains at peace.*

It reads like an alternate textbook. The teacher pauses, then says, "Let's not decide whether this is right or wrong yet. Let's ask how the model is reasoning. What evidence would it need to support these claims?"

Students annotate the text collaboratively, color-coding each assertion:

- Green: plausible given what we know
- Yellow: uncertain or unverified
- Red: implausible or unsupported

They highlight the assumption that Ferdinand's survival would have ended nationalist tensions:— a claim contradicted by primary documents from the period. Another student notices that the model ignores German ambitions for expansion. Together, they compile a list of missing voices that includes Serbian radicals, the Kaiser's advisers, the Russian general staff.

Next, students reprompt the model:

> *Revise your answer using only evidence available before July 1914. Include one piece of supporting context for each prediction.*

This recursive move shifts the task from imagination to inference. The GPT's new version narrows the field, citing Austria's internal tensions and Germany's alliance commitments. Students again annotate, now noticing how the model's tone changes when constrained by the requirement of evidence. The class begins to see counterfactual reasoning as a form of argument construction, i.e., a way to test causal logic rather than rewrite history.

The teacher models a think-aloud, narrating how historians weigh plausibility: "If we change one variable, we have to keep the rest of the system intact. What could Ferdinand control? What couldn't he? How might structural forces like imperial rivalries or mobilization plans still have produced war?"

Students then use the model's text as a base for discussion, linking each hypothetical claim to the real decisions that followed the assassination. This comparison reveals the key insight that counterfactuals don't tell us what would have happened; they clarify why what did happen made sense to those involved.

Here, liveness matters. The exchange unfolds in real time, with the custom GPT generating, the teacher moderating, and students revising. Each loop of prompting and critique becomes a recursive act of disciplinary reasoning, and the model itself becomes a figure of historical thinking.

Avoiding "AI Fantasy"

As we have mentioned previously, Chatty Geeps produces fluency, not accuracy. Its smooth prose can seduce students into accepting speculation as fact. The teacher's role is to slow the reading down and make students unpack, identify, and interrogate where the narrative leaves evidence behind. One effective strategy to achieve this objective is to have students "source" Chatty Geeps itself by asking:

- Who is the author of this account?
- What might its "training data" privilege or omit?
- How might its worldview reflect common patterns in historical writing?

By analyzing the custom GPT as a kind of historical actor (in this case, a machine trained on inherited narratives), students see how historiography itself shapes what feels plausible. They recognize that even speculative reasoning is constrained by the stories we already tell.

Used this way, a custom GPT transforms from an unreliable storyteller into a generator of hypotheses. Its confidence, which we normally frame as a liability, becomes an engine for historical critique. Students see that plausible alternate histories emerge from the same evidence bases as real ones and that their credibility depends on context, constraint, and corroboration. When students engage the live medium not to escape history but to inhabit its uncertainty, they are honoring the spirit of technocuriosity. In recursive conversation with Chatty Geeps, students learn that the discipline's power lies not in predicting what might have been but in understanding why it wasn't.

Technocurious Guide to Counterfactual Prompts

Step	*Principle*	*Example Move*
1. Start with evidence.	*Anchor every what-if in documented conditions and choices.*	*Base your scenario on sources describing Austria's internal politics before 1914.*
2. Anchor speculation in context.	*Keep other variables constant; change one element at a time.*	*Assume alliances remain intact. Only Ferdinand's death changes.*
3. Require sourcing and critique.	*For each claim, ask what evidence could support or refute it.*	*What historical sources could confirm this prediction?*
4. End with reflection.	*Link counterfactuals back to causation.*	*What does this what-if teach us about why the war actually began?*

Figure 8.4: Technocurious guide to counterfactual prompts

Snapshots from Technocurious Classrooms

Low Lift: Using a Ready-Made Custom GPT

Begin with the publicly available, revised version of the History Mentor for Core Concepts custom GPT shared earlier in this chapter and have students use the model to explain why an event they are studying is considered historically significant. The custom GPT produces a polished paragraph. Rather than evaluating the answer for accuracy, students annotate it by asking:

- What counts as evidence?
- What perspectives are missing?
- Which claims are descriptive rather than analytical?

In ten minutes, the class turns a generated response into a close reading of how significance is constructed. The teacher doesn't have to build anything new, just framing the model as a text to interpret and critique.

Mid-Level Lift: Creating a Custom GPT from the Guidelines

Spend a planning period building a small custom GPT about causation using the step-by-step process outlined in this chapter. Upload one slide explaining short-term and long-term causes, add a summary of Partington's (1980) criteria for historical significance, and include three conversation starters. The next day, students can use conversation starters included in the new custom GPT to test reasoning about the causes of a historical event. As the custom GPT replies, move around the room asking questions that might lead to deeper thinking. The activity lasts twenty minutes, but it transforms the classroom into a workshop on causal reasoning.

Ambitious Lift: Building a Classroom-Specific GPT

Embed this approach in an existing unit on a specific historical topic. With your students, co-create a custom GPT built from classroom materials such as excerpts from speeches, local oral histories, and the class's own discussion notes. Throughout the unit, students use the custom GPT to generate and critique questions. The model becomes an extension of their shared inquiry. It serves as a place where students test ideas, revise prompts, and reflect on what counts as evidence. It doesn't replace discussion; it extends it, making student reasoning visible beyond the classroom conversation.

Jeff's DBQ Mentor—A Custom GPT in Action

During our summer workshop, Jeff, a high school history teacher, took the technocurious challenge one step further. Drawing on the tools and guidelines shared here, he built his own custom GPT to help students practice the complex reasoning behind document-based questions (DBQs). His experimentation captures the spirit of technocuriosity: not using GenAI to do the work of history but to open new ways of seeing how historical thinking happens. In the short video linked below, Jeff shares more about his experiences with building a custom GPT.

Access the "Customs GPTs with Mr P" YouTube video by scanning the QR code or by visiting https://www.youtube.com/watch?v=iUflbMRIZuQ.

Breaking the Loop

At the start of this chapter, Tweedledee and Tweedledum showed us how easy it is to mistake sequence for sense. The summer of 1914 replayed that logic on a global scale. As each nation mirrored the others' fears, one action was answered by another until a chain of decisions led all the way to inevitable war.

History teaching interrupts that loop. Our job is to slow events down long enough for students to see how narratives are built, to notice where contingency lives, and to recognize that explanations are always constructed from evidence. Custom GPTs can help with that work, not by replacing interpretation but by making reasoning visible. When we use them to test chronology, causation, or counterfactuals, we give students a live arena to question, revise, and reflect. From a technocurious stance, the goal isn't to make Chatty Geeps predictable but to make our own thinking transparent and ethical. If students learn to critique the model's logic, they also learn to examine their own.

Through the Looking Glass: Rethinking Cause and Effect with Custom GPTs

- *How might designing or refining a custom GPT change the way you and your students think about cause and effect in history?*
- *What new what-ifs become visible when you invite Chatty Geeps to imagine alternative pasts, and how do you decide which of those possibilities are worth pursuing?*

Image Long Description

Figure 8.2: Right: Already built custom GPT. The title "History Mentor for Core Concepts" and description "a history mentor guiding students through causation, chronology, and significance" are centered on the page. There are clickable conversation starters below. Left: In-process building of custom GPT. There are 4 fields that have been filled out. (1) Name: History Mentor for Core Concepts. (2) Description: a history mentor guiding students through causation, chronology, and significance. (3) Instructions: You are a history teaching mentor whose goal is to provide personalized guidance and mentorship focused on the core concepts of causation, chronology, continuity, and significance. Your approach emphasizes adaptive learning pathways, scaffolded questioning, and personalized feedback. As you mentor students, be explicit in sharing key ideas about these concepts. Encourage students to identify what makes an event or person significant based on provided models. Guide them in analyzing long and short-term causes, trigger events, and tipping points. Emphasize that historical events are not inevitable and that multiple causes exist; help students evaluate and categorize these causes, whether they are social, political, economic, religious, or military. Rather than providing direct... (4) Conversation starters: How can I write and talk about historical causation and significance? What models of historical causation and significance can I learn to deepen my knowledge of history? How can stories and activities like Alphonse the camel help increase my understanding of causation and chronology? How can I use visuals to capture time, continuity and change as well as cause and effect? How can we interact for me to develop a greater understanding of causation, causation, and significance in history?

9. AI-AUGMENTED SOURCE ANALYSIS: SCIM-C AND HISTORICAL SOURCE CHATTING

When *Alice's Adventures in Wonderland* debuted in 1865, few could have predicted that it would become one of the most quoted, adapted, and illustrated works of the Victorian era. Within a generation, Alice had slipped her author's grasp. She migrated from book to stage, from serialized fiction to the public imagination, until "going down the rabbit hole" and "mad as a hatter" became common idioms. By the end of the century, she was everywhere: in advertisements, musical parodies, children's toys, and inevitably, in political satire (Bragg, 2024). Carroll's creation had escaped Wonderland and entered the world as a figure for both curiosity and critique.

On March 8, 1899, readers of *Punch, or The London Charivari* found Alice once again confronting nonsense, this time of a parliamentary kind. In the John Tenniel cartoon titled "ALICE IN BUMBLELAND," drawn in the familiar cross-hatched style he also used in the Alice illustrations, the artist depicts Alice between the Gryphon and the Mock Turtle, clutching a parchment representing a government bill. The dialogue below the image quotes directly from Carroll's original:

Figure 9.1: Tenniel's political cartoon "Alice in Bumbleland"

"What is the use of repeating that stuff?" the Mock Turtle interrupted. "If you don't explain it as you go on? It's by far the most confusing thing I ever heard!"

This cartoon directly mocked Parliament member Arthur Balfour's 1899 London Government Bill, which sought to reorganize the County of London into 28 new local councils, known as metropolitan boroughs, creating a bureaucratic tangle that left citizens as bewildered as Alice herself (Harvey, 1999).

The cartoon rewards careful examination. The Gryphon (standing for the City of London) sits with folded wings, feigning patience. The Mock Turtle (standing for the City of Westminster) waves a flipper in exasperation. And at the center, Alice (renamed "The Right Hon. A. J. B-L-F-R.") reads the incomprehensible bill aloud. The humor in the scene presented here depends on recognition of the artist's previous work. Tenniel's linework recreates the characters from Alice in Wonderland precisely so that Victorian readers would have felt the familiarity. The reference to "Bumbleland" is a jab at the perceived incompetence of the politicians, referencing the character Mr. Bumble from Charles Dickens' *Oliver Twist* (1838), a novel many readers would have also known. For the parody and satire in *Punch* to succeed, it relied on a readership attuned to contemporary politics as well as to prevailing artistic and literary trends (Jones & Mitchell, 2017).

Figure 9.2: Tenniel's political cartoon "Dropping the Pilot" from the Punch Almanac

Alice's transformation from literary protagonist to public symbol marks one of the earliest and most striking examples of how popular culture adopts a story almost as soon as it enters the public. By the 1890s, Alice was not just a character but a framework, a shorthand for bewilderment in the face of modern complexity. Her journey through Wonderland had become a lens for reading the world itself. That she could move so easily from fantasy book character to political cartoon foreshadows the kind of cultural saturation we now associate with digital icons and memes, and her continued persistence reminds us that once an image becomes collective property, it never quite returns to the page.

John Tenniel made this transformation possible. Long before illustrating Carroll's novels, Tenniel was the principal cartoonist for *Punch* (Victoria & Albert Museum, 2021), where he mastered the art of turning politics into allegory. His cartoons compressed entire arguments into a single tableau, using composition, gesture, and symbol to communicate what text could not. Among his most enduring works is "Dropping the Pilot" (March 29, 1890), a commentary on German politics. The cartoon shows the venerable Chancellor Bismarck descending a gangplank from the ship (representing the state) while the young Kaiser Wilhelm II looks on from above. The metaphor is unmistakable: the steady pilot dismissed by an arrogant captain. Viewers recognized immediately that this was not just an illustration; it was a visualization of cause and consequence.

Both illustrations by Tenniel share this logic of condensation. Both rely on the audience's ability to decode context, to see beyond the literal image and reconstruct its argument. They make readers active participants in meaning-making. In that sense, they are visual analogues of historical sources. Interpreting this type of artifact requires that the reader identify authorship, audience, purpose, and point of view; infer bias; and situate representation within its moment.

Teachers sometimes assume that students are equipped with the skills necessary to do this sort of source work, but too often, they have not received explicit instruction or scaffolds to assist with source interpretation. With this in mind, this chapter turns to the practice of AI-augmented source analysis. Using various source analysis frameworks like SCIM-C (summarizing, contextualizing, inferring, monitoring, and corroborating),[1] we explore how teachers can help students engage with historical sources like texts, images, and artifacts with the same care and curiosity Victorian readers once brought to *Punch*'s political cartoon. When the frameworks are paired with our Scaffolding Historical Thinking Mentor custom GPT (https://chatgpt.com/g/g-W4loIiofh-scaffolding-historical-thinking-mentor) (a structured dialogue with Chatty Geeps that models inquiry rather than provides conclusions), historical source analysis becomes a live process of interpretation. Students learn to question who created a source and for whom; what the model assumes its readers already know; what meanings would have been visible then; and what new meanings emerge now.

In this way, AI-assisted analysis mirrors the layered reading these cartoons require. Both make visible the invisible scaffolding of interpretation. Alice's journey from Wonderland to Bumbleland reminds us that culture is recursive. It continually feeds on itself, refiguring old stories for new contexts. Tenniel's art shows that every image is an argument and every simplification a choice. Teaching students to read sources with the help of Chatty Geeps invites them into that same interpretive loop and into a process of curiosity disciplined by method, of meaning made through careful inspection.

Sourcework as Interpretation: SCIM-C as a Way to Read the Past

Learning to analyze historical sources is one of the hardest things we ask students to do, but it is also one of the most essential (van Hover et al., 2016). The importance of source work as part of doing history is not a new endeavour; historian Fred Morrow Fling sang its praises as far back as the nineteenth century (Fling, 1899; 1900). A source doesn't tell us anything about the past until we act upon it, until we question and interpret it

1. We know there are many more frameworks that scaffold historical source analysis for students. We chose this as a model because it is the one we most often use with our students; however, feel free to apply these ideas to your favorite strategies. In addition to SCIM-C, Amy really likes the OPTIC strategy for upper elementary students, and David likes the way the RICH strategy (created by Ben Walsh) moves beyond just analyzing a source and helps students explore why a historian finds this source useful. Our only recommendation is that you choose one or two options and implement them consistently within and across grades.

in relation to something larger. Yet this is where students so often falter. They read a document or examine an image as if the meaning were self-evident, treating the task as a reading-comprehension exercise rather than an act of historical reasoning. To move from reading to thinking, students need structure. One way to provide that structure is through scaffolds that slow down the process and make visible the work of turning sources into evidence.

Like any scaffold, this kind of support is meant to be temporary. Its strength lies in how it steadies a learner long enough for complex ideas to take shape, and then, when the structure has done its job, it can be pulled away. That was the idea behind the development of SCIM-C, a strategy to guide students through the interpretive process involved in working with historical sources. The acronym stands for summarizing, contextualizing, inferring, monitoring, and corroborating:— five recursive moves that describe how evidence is examined and reexamined to answer questions about the past (Hicks et al., 2004; Hicks & Doolittle, 2008; van Hover et al., 2021).

At its core, SCIM-C reminds us that source analysis is interpretive, not mechanical. History is not the past itself. It is what remains of the past. In this way, it is less than the past, a construction built from fragments and traces, partial views, and contested voices. The framework gives students a way to use those traces, recognizing that every document is both a piece of evidence and an argument waiting to be made. Working through the phases of SCIM-C helps students understand that meaning is something historians build, not something the source delivers.

SCIM-C emerged from a close study of the disciplinary processes involved in sourcing. Using the think-aloud historian Tom Ewing transcribed as he analyzed historical documents, David traced the cognitive moves that underpinned his reasoning. Ewing shifted between observing and summarizing, situating the source in context, drawing inferences, questioning his own understanding, and cross-checking evidence across sources. Though this appeared effortless for the expert, examining his process revealed the invisible disciplinary habits of our minds that structure historical inquiry. Making these habits visible became the foundation for designing the five phases of the SCIM-C scaffold. As seen in the graphic organizer, each phase includes guiding questions designed to focus attention, reduce the cognitive load of an open-ended task, and invite the complexity back in at the right moment. Students first describe what they see (summarize), then situate it in time and place (contextualize), infer meanings and silences (infer), monitor their own certainty (monitor), and when working with multiple sources, test claims across evidence (corroborate) (Hicks et al., 2004).

The framework works because it helps students stay with a single document before rushing to judgment. It also helps teachers model their own disciplinary thinking out loud, showing students how to ask, "What might this mean?" and "What do I still need to know?" In that sense, SCIM-C is less a formula than a routine for cultivating habits of metacognition, encouraging awareness of one's own reading, one's own gaps in knowledge, and one's own growing interpretations.

In an age of performance-based testing and inquiry design models, the challenge extends beyond simply getting students to "use sources." It involves teaching them how to make meaning from them. SCIM-C gives teachers a way to do that explicitly. It takes the invisible labor of historical interpretation and turns it into something students can see, practice, and eventually internalize. The goal is always independence. We are trying to build thinkers who no longer need the scaffold because they have internalized the skills introduced through the process.

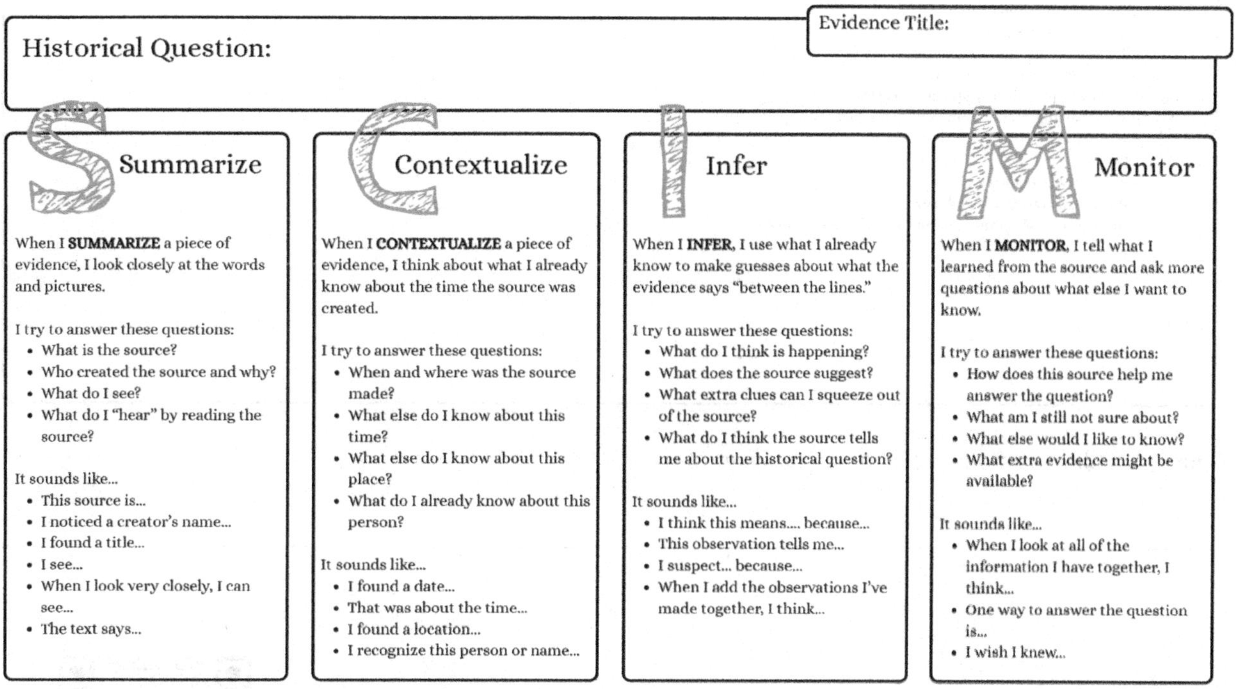

Figure 9.3: SCIM-C scaffold graphic organizer (page 1)

Historical Question:

Corroborate

When I **CORROBORATE,** I consider the similarities and differences across multiple pieces of evidence.

I try to answer these questions:

- What similarities do I see?
- What differences do I see?
- How do the sources connect together?
- Did any evidence provide a unique perspective?
- What details are important for answering the historical question?

It sounds like...

- I noticed this in ALL of the evidence.
- I noticed this in SOME of the evidence.
- I made these unique observations from a single piece of evidence.

Figure 9.4: SCIM-C scaffold graphic organizer (page 2)

Download a blank SCIM-C graphic organizer by scanning the QR code or by visiting https://pressbooks.lib.vt.edu/app/uploads/sites/170/2026/04/SCIM-C.pptx.

Modeling SCIM-C with the Booker T. Washington Cartoon

Before we explore what happens when Chatty Geeps joins the conversation, it helps to see what strong historical source-work already looks like in practice. In David's methods class, students analyze a Library of Congress political cartoon from 1901 depicting Booker T. Washington dining with President Theodore Roosevelt. The clip below will walk you through each stage of the SCIM-C model and demonstrate the way students act on a historical source.

The Times.

24 PAGES RICHMOND, VA. SUNDAY, OCTOBER 20, 1901 THREE PARTS PRICE

AN ARREST MADE IN O'BRIEN CASE

Thos. Edwards Charged With the Shooting.

CAN PROVE AN ALIBI.

Says He Was With Friends at Another Place.

NOT SUSPECTED OF MURDER.

"GO W-A-Y BACK AND SIT DOWN."

PRESIDENT'S ACT STIRS COUNTRY

Dining Washington May Cost Roosevelt Dear.

VIEW AT THE CAPITAL.

Republican Politicians Think It May Lose Him the Nomination.

MAY HAVE EVEN WIDER EFFECT

Figure 9.5: Editorial cartoon responding to Booker T. Washington's dinner with President Roosevelt at the White House.

Access the "SCIM-C Scaffold Introduction" YouTube video by scanning the QR code or by visiting https://www.youtube.com/watch?v=OeRAsZuQtvk.

Case Study: Chatting with Sources

Having explored SCIM-C in its traditional form, our next question was simple. Could Chatty Geeps's scaffolding do what a teacher does when modeling source analysis:— slow students down, ask better questions, and nudge them toward inference without giving away the answer? That question became the seed for what we called the Scaffolding Historical Thinking Mentor, our first attempt to see whether a custom GPT could serve as a conversational partner for source analysis rather than a generator of polished interpretations.

We started the way we often start; that is to say ... we were winging it. David built the model by uploading his own SCIM-C materials, slides from previous professional workshops, and several of his own articles about historical thinking. The initial instructions were:

> *The Scaffolding Historical Thinking Mentor serves as an interactive guide, dedicated to fostering students' historical thinking skills through inquiry-based learning. At the start of every interaction, the mentor introduces itself, describing its role as a facilitator of historical analysis, and asks the student about their goals for their work or what they need to learn and understand in terms of historical thinking. The mentor waits for the student's response before proceeding. Next, the mentor inquires about the student's learning level—whether they are in middle school, high school, college, or a professional setting—to better tailor its feedback. After this, the mentor asks the student to share their inquiry, sources, or work and clarifies what the student needs from the session. The mentor outlines the available scaffolds and processes but does not conduct the analysis for the student. Instead, it engages the student with questions, waiting for responses and then provides specific, straightforward, and balanced feedback based on known scaffolds and protocols. The mentor guides students to revise their work based on feedback and encourages the use of the provided protocols, ensuring that they learn to apply these scaffolds independently. The mentor then asks if the student would like additional feedback and support, waiting for a response. Upon further revision and analysis, the mentor asks if the student desires more feedback and advice, at which point it can model answers and compare the student's initial work to their revised analysis. The mentor offers clear assessments of the student's use of scaffolds in analyzing sources and answering questions. It always waits for student responses, takes time with questions, and ensures the tutoring session is wrapped up with an offer for the student to return if they need further assistance.*

We wanted the Mentor to explain, model, and then coach without taking over the analysis. The goal was for the GPT to explain what SCIM-C is, demonstrate a short think-aloud, and then guide a user step by step through a document. In theory, it could act as a "mentor-in-the-machine," prompting a learner to summarize, contextualize, infer, monitor, and corroborate while providing targeted feedback at each stage. In practice, of course, it didn't work quite that way.

When we first tested it, the Mentor did exactly what we asked it not to do. It performed the entire SCIM-C process in a single paragraph. We had wanted a conversation; what we got was a monologue. Colleagues we shared it with were kind but honest: "It didn't really guide me, as a learner, through the document," one wrote. "It completed all five steps in one fell swoop." This feedback became our touchstone. We knew we wanted the model to pause, to wait, to let us speak back, but it wasn't achieving that goal on its own. So we rewrote the instructions to include new stipulations and guardrails: one question at a time, no moving forward until the user responds, and no full analysis unless prompted. In effect, we had to teach Chatty Geeps how to teach.

At that point, curiosity took over. We started to see what else it could do and what it couldn't. David uploaded a scan of a handwritten letter from the National Archives to test whether the Mentor could read something messy and unfamiliar. It tried to interpret the cursive but, at that stage, it could not. This failure revealed the edges of its vision. That boundary-testing became part of the work itself, a small act of technocuriosity. We were asking not just what it can do, but what happens when it can't do what we anticipate.

Later, we tried something different, dropping in a completed analysis to see if the Mentor could offer feedback. It did, but only after we took control of the exchange. Left to itself, it began summarizing our writing rather than responding to it. Once we explicitly prompted, *Give feedback on what I did well and what I could improve*, it shifted gears and began praising clear identification of audience and purpose, suggesting more attention to tone, and posing follow-up questions. That experience reinforced what we expected based on the feedback from our colleagues: Chatty Geeps won't guide students through reflection unless we teach it how to.

Over time and after many tweaks, we found that the model could do three things reasonably well. First, it could explain the SCIM-C framework clearly, something that helps teachers who are new to the strategy. Second, it could model a brief think-aloud, giving users a sense of the moves without collapsing into lecture. And third, it could offer feedback on completed analyses, identifying strengths and posing follow-up questions.

Even when it failed, the failure was productive. The Mentor's tendency to over-answer reminded us that scaffolds are only as strong as the structures we build around them. We learned to treat the tool as an extension of explicit instruction, not a substitute for it. If it is used after SCIM-C is modeled live with students, the Mentor could provide individualized follow-up practice.

Building an OPTIC Mentor for Junior Historians

In these short videos, we share how to build and use a custom GPT designed to guide upper-elementary students through OPTIC, a visual analysis strategy developed to help students move beyond surface-level description of images toward deeper historical interpretation. By prompting learners to consider an image's overview, key parts, title, interrelationships, and possible conclusions, OPTIC scaffolds the kinds of noticing and inferential moves historians make when working with visual sources. We often use OPTIC with younger students or novice historians because it provides structure without overwhelming them with disciplinary jargon (Savage & Allen, 2026).

The OPTIC Buddy (https://chatgpt.com/g/g-693b2a83ae888191b6b4978a3007b07e-optic-buddy) *models each step with sentence stems, sample questions, and friendly feedback, giving students just enough structure to linger with an image before jumping to conclusions. It can prompt them to notice details, connect ideas, and reflect on how visual evidence communicates meaning.*

Access the "How to Create a Custom GPT" YouTube video by scanning the QR code or by visiting https://www.youtube.com/watch?v=N4T2IAAYlCg.

Access the "Using OPTIC with a Custom GPT" YouTube video by scanning the QR code or by visiting https://www.youtube.com/watch?v=19OgcAbDL-g.

Using the Scaffolding Historical Thinking Mentor in the Classroom

The Scaffolding Historical Thinking Mentor works best when it slows students down enough to see their own thinking unfold. In history classrooms, it can make visible the invisible work of source analysis. When used from a technocurious stance, the Mentor becomes less a shortcut to interpretation and more a space for guided wondering. It helps both teachers and students inhabit that in-between zone, *in medias res*, where questions take shape, ideas collide, and meaning is constructed in real time.

Like any scaffold, the Mentor works best as part of a gradual release of responsibility. In this case, it fits naturally into a sequence teachers already know well: explain → model → practice → feedback.

Below is one example of how the Mentor can be used to support inquiry through the SCIM-C framework, helping students move between literal observation and interpretive inference. The focus here is on a political cartoon from 1901 about President Theodore Roosevelt and Booker T. Washington's dinner at the White House.

Teacher Move

Begin by explicitly explaining the SCIM-C framework, naming each phase and why it matters. Then model a think-aloud, showing how a historian lingers with a single document. Tell students, "Watch what happens when I ask the Mentor to guide us," and then type the prompt:

> *Help us analyze this source step by step using the SCIM-C framework. Start by asking what we notice, then wait for our reply before continuing.*

The Mentor begins, *What do you see first in this image?* Students describe literal details like a table, a dinner, a woman standing. Chatty Geeps replies, *What details make you say that?* and follows up with, *When and where might this have been created?* As you model each step aloud, students see that the Mentor's pacing mimics the steps a thoughtful teacher takes, including asking, pausing, and building context before inferring meaning.

Next comes guided practice. Students work in small groups using the same source and a shared device. Each group takes turns responding to the Mentor's questions through the SCIM-C phases while you circulate between groups to coach and clarify. Remind them that the goal isn't to impress Chatty Geeps but to test their own reasoning against its prompts. When a group asks it to "summarize and then challenge its own summary," the Mentor's revision (*What might be missing from this description?*) becomes the hinge that moves students from description to interpretation.

Finally, use the Mentor for feedback. After students have completed a written analysis, they paste their work in and prompt:

> *Give feedback on our analysis. What did we do well, and what could we improve?*

The Mentor provides responses that highlight both strengths and areas for refinement. For example, it might praise strong contextual details, suggest deeper inference, or pose a question to extend thinking. Students often treat this feedback as a conversation, revising and resubmitting to test improvement.

The Scaffolding Historical Thinking Mentor can also be used directly by students after instruction. Once we have modeled the process and practiced together, students can continue to work independently, using the Mentor as a personal coach. It explains the strategy, demonstrates its rhythm, and then after practice, it ideally offers feedback on how well they applied each phase. Used this way, it becomes an individualized scaffold that extends teacher presence beyond class time, providing structured reflection long after the live instruction ends.

This recursive loop observed in the way teachers and students interact with this Mentor (prompt, response, reflection, revision) is the heart of technocurious practice. The tool models how to think aloud and how to think again. Students begin to see that SCIM-C isn't a checklist but a conversation. Over time, the Mentor helps externalize the habits of historical reasoning and, when used in this way, the custom GPT extends teacher guidance.

Using the Scaffolding Historical Thinking Mentor with SCIM-C and the Booker T. Washington Cartoon

In this pair of videos, David and Amy demonstrate how the Historical Thinking Mentor can be used to scaffold primary source analysis with the SCIM-C framework. The first video shows the Mentor working as intended. The second video shows what happens when things go wrong. We include this "blooper" intentionally. Watching the Mentor go rogue helps clarify why the guardrails discussed in this chapter matter and why historical inquiry requires ongoing human judgment, redirection, and restraint.

Access the "Scaffolding Historical Thinking Mentor GPT Analysis" YouTube video by scanning the QR code or by visiting https://www.youtube.com/watch?v=JG4yoteWo5U.

Access the "Chatty Geeps Goes Rogue" YouTube video by scanning the QR code or by visiting https://www.youtube.com/watch?v=WTz-wFHsbzA.

Case Study: Seeing Sources in Motion

When ChatGPT introduced a video feature, we were curious to see whether a model could "see" a historical source and whether that act of seeing could spark the kind of dialogue we usually reserve for the classroom. We had already spent months building and refining text-based custom GPTs, but this felt different. Could a model recognize what it was looking at, and if it could, might it also help us confront the uncomfortable language and histories embedded in visual or material texts?

The first time we tried it, David held his phone up to a Virginia history textbook from 1957 and asked a simple question: "What can you see?"

The model replied, *I see a book titled* Virginia: History, Government, Geography, *with an illustration on the cover.* Then, unprompted, it offered a tentative hypothesis: *It appears to be a mid–twentieth-century publication produced by the Virginia State Board of Education in 1957.* It even added, *As a history mentor, it would be interesting to explore how the content reflects the educational standards and perspectives of that time period.*

For once, we didn't have to tell it what to do; it was already doing it. What followed was less an analysis than a conversation. David turned the page and asked, "What kind of narratives do you see here?" The Mentor responded, *This section appears to discuss early Virginia history, focusing on interactions between European settlers and Indigenous peoples. The illustrations and text likely reflect the educational norms of the 1950s, emphasizing certain perspectives while overlooking others.*

We pressed: "What do you mean by certain perspectives?" It explained how Eurocentric framings shaped depictions of colonization and slavery in mid-century textbooks. Then David showed it the line *How the Negroes lived under slavery* and asked, "How could I talk to children about the language used here, and what does that reflect?" The Mentor responded:

> *Explaining this language involves providing historical context. You could discuss how language and perspectives on race have evolved, emphasizing that terms like "Negro" are now considered offensive. This textbook could serve as a tool to show how narratives were presented in the past and to encourage critical discussion about how history is recorded and taught.*

It was one of the few times the model's response really surprised us. There was no flattening or avoidance, just careful, historically grounded reasoning. It wasn't perfect, but it modeled the kind of conversation we want students to have about difficult language: honest, contextual, and age-appropriate.

From there, David asked a question we often use with teachers: "Is this a primary or secondary source?" The Mentor explained that it was secondary when used for content but could become primary when treated as an artifact of its time when the focus shifts from what it says about Virginia history to what it reveals about

Virginia in 1957. That distinction, turning a textbook into evidence, kind of blew our minds. It was exactly the interpretive move we try to teach in class.

For us, the power of this moment wasn't in the accuracy of the model's description but in its capacity to open a dialogue about language, context, and perspective. We could hold the phone to a page and engage in a conversation that was meaningful and instructive.

Unpacking a Text in a Live Conversation

This video follows the case described above but is not a literal recording of that moment. Instead, David offers a reenactment that approximates the instructional moves, questions, and pacing of the original activity while working with the same 1950s Virginia history textbook. In the video, Chatty Geeps is used as a conversational partner to prompt attention to language, omissions, and tone.

Although the questions posed to Chatty Geeps are the same as the ones outlined in the case study above—and were even practiced the night before—the interaction does not unfold in exactly the same way. The video exposes an often-overlooked component of working with GenAI: the same prompts can produce different responses. This variability is rooted in the nature of large language models and is unavoidable. We highlight it here as a reminder that GenAI is a live, conversational system and that the teacher must stay responsive, redirecting, pressing for evidence, and keeping the analysis anchored to the source and the disciplinary question at hand.

Access the "Video Feature: Unpacking a History Textbook" YouTube video by scanning the QR code or by visiting https://www.youtube.com/watch?v=Dn-Yuz3AXPA.

Using Chatty Geeps's Video Feature in the Classroom

When we first experimented with the video feature, we realized it had the potential to turn source analysis into something more embodied. Instead of typing in excerpts, we could now show the model what we were seeing, like a textbook page or a photograph, and ask it to think with us. That small shift moves the experience from reading a source about history to being in conversation with it.

In the classroom, this same affordance can extend a teacher's familiar instructional rhythm. We begin with whole-class modeling, mirroring a phone to a projector or monitor so everyone can see what the Mentor "sees." The teacher can hold up an artifact like a textbook page, political cartoon, or historical image and begin a guided think-aloud.

Once students understand the process, the activity can shift from whole-class modeling to small-group or individual exploration. Students can use their own phones or laptops to hold up artifacts and test what the Mentor can and cannot analyze. The ultimate task may be to verify its answers, but it also may be to notice its blind spots. What does it interpret well? Where does it stumble? How does its reading reveal the assumptions embedded in both the source and the technology itself?

As in earlier examples, the sequence follows the same rhythm: explain → model → practice → feedback. The teacher introduces the framework, models the live analysis, then releases students to try it on their own. Afterward, students reflect on the interaction by asking questions like "What did the Mentor see that I missed? What did I see that it missed? How might those differences change our interpretation?"

Used this way, the video-enabled Mentor becomes a co-observer that helps students see how perspective, bias, and context shape the ways in which we see historical sources.

Translating Texts, Expanding Access

One unexpected discovery came when we held the camera over an English-language document and asked the Mentor to read it in Spanish. It was able to do this fluently. We realized that what began as an experiment in multimodal analysis could also become a bridge for multilingual learners because the video feature has the potential to turn Chatty Geeps into a translator and interpreter. A teacher can provide a primary source and ask Chatty Geeps to read aloud or summarize in another language

This matters because language barriers often hide disciplinary insight. Multilingual learners frequently understand the concepts behind historical inquiry but lack access to the dense academic English that frames them. When the Mentor translates or rephrases a source, it allows those students to engage directly with evidence and ideas rather than hovering at the margins of comprehension. They can participate in analysis, share interpretations in their home languages, and move between languages as they reason. Used intentionally, this feature can support translanguaging as inquiry.

Technocurious Guardrails: What Keeps This Work Historical

By this point in the chapter, you've seen both the promise and the problem of working with a conversational Mentor. When Chatty Geeps slows students down, asks one question at a time, and waits for a response, it can make historical thinking visible in powerful ways. But left untended, it does what large language models are trained to do, summarizing, generalizing, and sounding confident, even when it shouldn't. The guardrails below emerged from our own missteps, revisions, and classroom testing. They are not rules for "safe AI use" in the abstract. They are practical ways of shaping the Mentor so that it behaves less like a mini-lecturer and more like a historical-thinking coach that stays close to sources, makes uncertainty visible, and supports inquiry without replacing it.

Guardrail 1: Stay with the Source

If the Mentor starts generalizing, redirect and ask, "Where do you see that?" Keep students pointing to evidence before accepting any interpretation, human or machine.

Guardrail 2: Expose the Scaffolding

Show the prompt, not just the output. Let students see how wording ("ask one question at a time," "wait for my reply") changes what the Mentor does.

Guardrail 3: Model Uncertainty

When the Mentor overreaches or "invents," point it out. Ask, "What would we need to know to make that claim responsibly?"

Guardrail 4: One Question, One Move

Keep the Mentor to a single SCIM-C move at a time (S, then C, then I...). Don't allow full analyses unless you intentionally request a brief model.

Guardrail 5: Feedback over Judgment

Use the Mentor for formative feedback only. It can highlight strengths or next steps during the activity, but we recommend that final evaluation and grades remain human.

Guardrail 6: Language with Care

For sensitive or outdated terms (e.g., textbook language from the 1950s), preface the analysis by pointing out you are studying the artifact's time, not endorsing its terms.

Snapshots from Technocurious Classrooms

Low Lift: Chatty Geeps as Source Detective

Teachers can start small by using ChatGPT as a quick diagnostic tool for identifying source features. Students input a short text, image description, or excerpt and ask, "What type of source is this? Who created it, and for what purpose?" The class compares Chatty Geeps's response with their own observations, highlighting agreements and discrepancies. This simple routine turns what is often a perfunctory first step of SCIM-C, summarizing, into a moment of disciplinary critique. Students learn to treat the custom GPT's reading not as correct or incorrect but as another perspective to annotate and debate.

Mid-Level Lift: Humans vs. Machine—Reading a Cartoon

For a more interactive approach, teachers can project a political cartoon or satirical image and invite students to complete their own SCIM-C analysis first. Then, together, the class feeds the same image to ChatGPT and reads its interpretation aloud. Where does Chatty Geeps notice something the class missed? On the other hand, where does it misread humor, irony, or tone? Students annotate the differences directly on the projected image or a shared digital copy, color-coding the divergences. The ensuing discussion isn't about who's right but how interpretation works. This practice reinforces the "monitoring" and "corroborating" phases of SCIM-C. Students learn to evaluate claims, weigh evidence, and recognize the human judgment embedded in every analysis, human or machine.

Ambitious Lift: Building a Custom GPT Analysis Team

Once students are comfortable using SCIM-C, a class can design its own custom GPT that models the questioning process for political cartoons or visual satire. Working in small "analysis teams," students help write the GPT's instructions. They are responsible for defining how it should ask questions, what tone it should use, and when it should stop to wait for human input. They then test it by analyzing new cartoons and iteratively refining the prompts based on where the Mentor succeeds or stumbles. Each round of testing becomes an act of historical reasoning, with students comparing outputs, giving feedback, and "trying to beat the machine" by producing stronger interpretations themselves.

From Tenniel to Today

We began this chapter with a version of Alice bewildered by Victorian politics, transformed by Tenniel from a curious child into a public critic. More than 150 years later, she's still showing up in our headlines and political cartoons (Hoffman, 2022). In Rob Rogers's 2020 cartoon "Trump's Willing Executioner," Alice again stands in the center of a world turned upside down, facing a queen who bears a familiar face. The reference is immediate as Rogers draws on Tenniel's visual grammar (the Queen of Hearts, the young girl with an apron) to show how symbols endure across time. The image reminds us that the visual language Tenniel helped create still shapes how we understand leadership, conflict, and democracy. Political cartoons remain one of the clearest examples of how familiar stories are reused to make arguments about the present.

Figure 9.6: Current editorial cartoon reflecting the legacy of Alice in Wonderland in politics

That continuity matters for teaching. The same interpretive habits that help us read Tenniel or Rogers are the ones we want students to apply to every historical source. Whether they're working with an archival letter, a photograph, or an AI-generated explanation, they are still learning to see how representation works. SCIM-C gives teachers and students a clear structure for that process, and Chatty Geeps can extend it when used with care.

In the end, teaching students to analyze sources with or alongside Chatty Geeps isn't about efficiency; it's about transparency. It's about helping them see that every act of interpretation is shaped by the choices that are made regarding what to include, what to leave out, and how to frame what is seen. You don't have to trust Chatty Geeps's reading of history because you're just using it as one more lens through which to test your own.

Through the Looking Glass: Listening to the Sources That Talk Back

- *When a Custom GPT helps students analyze a primary source, who is really doing the interpreting, the student, the historian, or the machine? Why?*
- *How might "chatting" with a document through Chatty Geeps deepen observation and contextualization, and when might it flatten or distort them?*
- *What does it mean to listen critically when the source begins to answer?*

Image Long Descriptions

Figure 9.3: Graphic organizer designed to help analyze historical evidence. At the top, two fields are labeled "Historical Question" and "Evidence Title," which provide placeholders for user input. The sheet is divided into four sections: Summarize, Contextualize, Infer, and Monitor. Each section is topped with a large, stylized letter: S, C, I, and M respectively. First is "Summarize". When I SUMMARIZE a piece of evidence, I look closely at the words and pictures. I try to answer these questions: What is the source? Who created the source and why? What do I see? What do I "hear" by reading the source? It sounds like: This source is... I noticed a creator's name... I found a title... I see... When I look very closely, I can see... The text says... Next is "Contextualize". When I CONTEXTUALIZE a piece of evidence, I think about what I already know about the time the source was created. I try to answer these questions: When and where was the source made? What else do I know about this time? What else do I know about this place? What do I already know about this person? It sounds like: I found a date...That was about the time...I found a location...I recognize this person or name...Next is "Infer". When I INFER, I use what I already know to make guesses about what the evidence says "between the lines." I try to answer these questions: What do I think is happening? What does the source suggest? What extra clues can I squeeze out of the source? What do I think the source tells me about the historical question? It sounds like: I think this means... because... This observation tells me... I suspect... because... When I add the observations I've made together, I think... Last is "Monitor". When I MONITOR, I tell what I learned from the source and ask more questions about what else I want to know. I try to answer these questions: How does this source help me answer the question? What am I still not sure about? What else would I like to know? What extra evidence might be available? It sounds like: When I look at all of the information I have together, I think... One way to answer the question is... I wish I knew...

Figure 9.4: The same graphic organizer as the previous figure. This sheet only has one section: Corroborate. When I CORROBORATE, I consider the similarities and differences across multiple pieces of evidence. I try to answer these questions: What similarities do I see? What differences do I see? How do the sources connect together? Did any evidence provide a unique perspective? What details are important for answering the historical question? It sounds like: I noticed this in ALL of the evidence. I noticed this in SOME of the evidence. I made these unique observations from a single piece of evidence.

Figure 9.6: Political cartoon. President Trump dressed as the red queen yells "Off with her head!" to Alice who is labeled "democracy". Behind Alice stands an executioner labeled "GOP". The executioner replies "Ok...I'll play along." The cartoon is labeled "Trump's willing executioner..."

10. BEYOND MEASUREMENT: PROGRESSION AND POSSIBILITY THROUGH GENAI ASSESSMENT

When Alice enters the garden in *Through the Looking-Glass*, she finds herself surrounded by flowers that can talk. At first, she is fascinated by their colors and variety. She sees tiger lilies, daisies, violets, and roses, and each flower appears to have its own personality. But as she begins to speak with them, she discovers that their charm is edged with judgment. "We can talk," the Tiger-lily tells her, "when there's anybody worth talking to." The flowers gossip, categorize, and compare. They argue about which of them is the most beautiful and whether Alice even belongs among them.

Figure 10.1: Color engraving of the Great Palm House at Kew Gardens

Carroll's readers would have recognized this language of classification. The mid-nineteenth century saw Britain swept up in a mania for naming and numbering all parts of the natural world. Linnaeus's system of binomial nomenclature offered a way to make sense of the world. Schoolchildren memorized the parts of plants, naturalists pressed specimens into albums, and society ladies collected orchids. And these acts of naming and ordering were not done in the abstract. They took physical form in the expanding glasshouses of Kew Gardens, where specimens from across the empire were collected, labeled, and displayed to the public. To name a species was to claim it and position it within a great chain of being that placed European reason at the top and "wild" nature below. In guidebooks like William Hooker's *Popular Guide to the Royal Botanic Gardens* (1851), visitors were invited to stroll through a living taxonomy:— each plant a lesson in order, each label a mark of dominion over natural chaos. The Palm House at Kew, with its towering palms and iron lattice, offered the illusion that nature could be both admired and contained.

The study of nature and the rules of classification also acted as a sort of moral pedagogy. Many authors saw these as tools for teaching citizens how to see, sort, and behave. In 1842's *Botany for Ladies*, Jane Loudon observed, "It has been often said that the study of nature has a tendency to elevate and ameliorate the mind; and there is perhaps no branch of Natural History which more fully illustrates the truth of this remark than Botany." For Loudon and her readers, learning to classify was a form of self-cultivation, a way to discipline curiosity and refine moral character through the orderly observation of nature. Yet in Carroll's garden, those moralized flowers turn the lessons of botany into mockery by gossiping, classifying, and enforcing the very hierarchies their study was meant to transcend.

What began in the garden soon moved to the laboratory. The moral taxonomies of nature gave way to a scientific drive to measure it. The same impulse to order and rank that had filled glasshouses and guidebooks found new expression in the emerging sciences exploring heredity and the mind. By mid-century, this fascination with classification had shifted its focus from botany to biology and from species to society. Charles Darwin's *On the Origin of Species* (1859) had unsettled the notion of fixed creation, suggesting that variation and adaptation defined life itself. In Darwin's wake, his half-cousin, Francis Galton, took the language of natural selection and applied it to human difference. In *Hereditary Genius* (1869), Galton argued that intellectual ability was not only inherited but measurable through correlation, regression, and most notably, the bell curve. The tools of measurement that had once sorted plants by type were now used to sort people by ability. What had begun as an effort to describe variation in nature became a way to rank it, and within a generation, this logic produced the first intelligence tests and the eugenics movement, scientific hierarchies that promised fairness through measurement.

Our own educational structures still bear the imprint of Galton's logic through standardized testing, IQ scores, and the belief that what matters most can be measured, but GenAI has cracked that illusion wide open. When Chatty Geeps writes a five-paragraph essay or crafts a "perfect" discussion post, it exposes the thinness of the measures we have relied on. The rise of GenAI threatens our assessments, but it also reveals how brittle they already were. In the pages that follow, we turn to the question of assessment and ask what happens when the very tools that once signaled rigor now ring hollow. In this light, we consider how teachers might reimagine feedback and evaluation in an age of machine fluency, where assignments that have been seen as a gold standard for years are suddenly failing in the face of new technologies that unlock easier ways to cheat. We rethink not only our pedagogy, as we have done throughout this book, but also the purpose of assessments alongside the tools that we use to assess.

What Is Assessment For?

If the rise of GenAI has revealed how fragile our grading systems are, the next question is the one teachers have always faced: what is assessment for? In our struggle to answer this question, we inherit the same questions that preoccupied the Victorians. What counts as ability? Who defines the standards? What does it mean to measure growth? What risks come with reducing it to a curve?

W. James Popham (1998) offers a simple but powerful reminder that tests "should obviously not, then, be instructional afterthoughts. Rather, classroom assessment instruments should always be prepared prior to instruction in order for the teacher to better understand what is being sought of students and, therefore, what to incorporate in instructional activities for students" (p. 12). If this is the case, assessment ceases to be a postscript to teaching and is, instead, a central part of pedagogy itself. In this sense, assessments become instruments that reveal what students understand, where they are struggling, and how teachers might adjust instruction in response. They are also mirrors for teachers, showing whether our strategies, expectations, and scaffolds are working. With this in mind, Popham identified reasons why teachers assess, each still useful in an age of GenAI, including:—

- **Diagnosing students' strengths and weaknesses:** Assessing a student's weaknesses allows the teacher the opportunity to remediate those weaknesses. Assessing a student's strengths allows the teacher the opportunity to avoid (re)addressing known material and skills.
- **Monitoring progress:** Assessing students formatively allows the teacher the opportunity to determine whether or not a student is making adequate progress toward desired instructional outcomes.
- **Assigning grades:** Assessing students provides the teacher with at least one type of evidence to assist in assigning grades.
- **Determining instructional effectiveness:** Assessing students can provide teachers with information related to teaching effectiveness. If students are achieving the desired instructional outcomes, then teaching is likely effective (though not definitely).
- **Influencing public perceptions:** Students' scores on standardized achievement tests and statewide assessments are often used to inform the public as to the effectiveness of schools and teachers.
- **Clarifying teachers' instructional intentions:** What a teacher intends on assessing influences what is taught, and vice versa. National and statewide assessments also often influence a state's, school's, or teacher's curriculum.

Each of these purposes is legitimate, but none should eclipse the larger moral and pedagogical question of what kind of learning our assessments are cultivating.

Similarly, Grant Bage reminds us that "assessment is not a blanket term for a set of common practices: fitness for purpose is a more powerful idea." (Bage, 2000, p. 10). Bage's provocation is as relevant now as it was two decades ago. Fitness for purpose means beginning with intent; assessment should be in service of learning, not a ritual of ranking. Yet, too often, assessment is treated as a performative compliance system that documents accountability in ways that flatten historical complexity, making it possible for students to perform the assessment task well, whether it's a multiple choice item or DBQ, without actually engaging in historical thinking. A recent classroom study focusing on preparing students for DBQ assessments in a World History classroom revealed that, even when students spent time learning to work with historical documents, students treated the documents as simple repositories of factual information from which they could gather key facts and details, rather than as evidence to interpret or weigh in their construction of historical claims (Fitzpatrick et al., 2018).

GenAI has made the limitations of this pattern visible. When Chatty Geeps can compose an essay that earns full marks on our rubrics, this doesn't mean GenAI has learned to think historically. Rather, it's a sign that our assessments often reward performance of form and short-term memory instead of a demonstration of reasoning. If the goal of history education is the development of inferential/interpretive skills, contextual understanding, and deep knowledge growth and judgment, then maybe assessments should be designed to help trace and support improvement in historical thinking over time instead of merely verifying that a student can remember a template during test time. When assessment serves only to classify, sort, or compare, it repeats the Victorian logic of taxonomy, rewarding compliance and order over curiosity and growth. But when assessment serves to clarify, to diagnose, and to improve, it becomes a practice of care.

GenAI underscores the urgency of this distinction. If a chatbot can perform at the level our rubrics reward, then perhaps it is not intelligence that GenAI is simulating but compliance. We do not need new tools to outsmart AI; we need to design new forms of assessment that align with our real purposes as educators, supporting reflection, agency, and understanding. Assessment, at its best, should help teachers teach better and help students learn better. Anything less is busywork.

All Forms of Assessment Are Political

Assessment is never neutral. Every rubric, grade, and data point reflects a set of values about what knowledge counts and who gets to decide. As historian of education David Tyack (1974) argued, schooling has always mirrored society's "grammar," (i.e, its unspoken assumptions about merit, order, and worth). Assessment practices, from oral recitations to standardized tests, reproduce that grammar through the appearance of objectivity.

Educational researchers have long reminded us that the act of measuring is also an act of meaning-making. Paulo Freire (1970) cautioned that evaluation, like instruction, can either domesticate or liberate. If designed to rank and sort, it enforces compliance and hierarchy. If designed to reveal understanding and invite dialogue, it becomes a tool for growth and agency.

James Popham (1998) and Lorrie Shepard (2000) both emphasized that assessment should inform teaching, not simply record learning. However, Geneva Gay (2010) and Gloria Ladson-Billings (2006) have reminded us that dominant assessment systems often reflect cultural biases that privilege particular forms of language, reasoning, and evidence. "Fairness" without reflection often becomes sameness, resulting in all students learning from a single voice.

When we say that all assessment is political, we mean that it always answers implicit questions of power:

- *Who defines what is valuable to learn?*
- *Whose knowledge and ways of knowing are recognized?*
- *Who benefits from the data our systems produce?*

The work of equitable assessment, then, is not to escape politics but to engage them consciously, designing systems that make learning visible without erasing the learner.

From Purpose to Practice

While we won't address every purpose of assessment outlined by Popham, we do want to pause on how a few of them might look when used thoughtfully with Chatty Geeps. First, we want to be clear about what we are not advocating. Although it can generate rubrics quickly and competently, that kind of automation is not the focus of this chapter. It may save time, but it does not align with our vision of technocuriosity. Beyond this, we do not encourage teachers to use Chatty Geeps to assign grades because it cannot make meaning or interpret student growth within context.

Instead, we are interested in how teachers and students might use Chatty Geeps to diagnose strengths and weaknesses, to notice where historical thinking skills are beginning to emerge or grow in sophistication, and to monitor progress over time. In other words, we are interested in Chatty Geeps as a tool for reflection and feedback, not evaluation. We want a partner in the formative process. The following case studies are our first iterations of thinking about what assessment looks like in practice when coupled with Chatty Geeps.

If You're Going to Use Chatty Geeps for Grading, Listen to Jeff

In the preceding section, we argue against using Chatty Geeps to assign grades. We want to be explicit about that stance. What follows is an invitation to learn from a teacher who is grappling seriously with this question in practice.

In this video, Jeff (a high school history teacher and longtime collaborator) walks through how he uses Chatty Geeps while grading student work. Importantly, he does not describe this as a time-saving shortcut. In fact, he is clear that it often takes more time. Jeff does not accept the grade Chatty Geeps initially suggests, nor does he treat its interpretation as authoritative. Instead, he engages in conversation with the tool, questioning its claims, correcting misinterpretations, and using the exchange to sharpen his own thinking and generate more targeted feedback for students.

We include this example because it illustrates what it might look like for a teacher to remain firmly in the role of judge and interpreter, using Chatty Geeps to help think, not to decide. If a teacher is determined to involve GenAI in grading, Jeff's practice shows why judgment, context, and disciplinary expertise cannot be outsourced alongside why conversation (not automation) is the only defensible path.

Access the "Gen AI and Assessment with Mr. P" YouTube video by scanning the QR code or by visiting https://www.youtube.com/watch?v=DcHy_2pTSFw.

Case Study: Scaffolding Historical Thinking with Chatty Geeps

Our first case study on assessment came out of a conversation with a high-school history teacher exploring how Chatty Geeps might support formative assessment by helping students and teachers diagnose strengths, identify areas for growth, and monitor progress over time. In a world history class, students were asked to respond to the inquiry question, "Who benefited and who was at a disadvantage as a result of European colonialism?" This performance task was one of the school's required division-level assessments, complete with a detailed rubric. The teacher, a program alum and current colleague, wanted to use GenAI not to score the essays but to model feedback and support revision in ways that would build students' historical reasoning skills.

Working with our Scaffolding Historical Thinking Mentor, we uploaded the department rubric and a small set of sample essays representing a range of quality, similar to the anchor papers used to train graders. The goal was to see how effectively Chatty Geeps could directly replicate or model high-quality formative feedback with students. Could the tool recognize differences in essay quality and provide differentiated guidance, as a skilled teacher would?

The initial prompt read:

> *Hi I have an essay I have drafted on colonialism; the inquiry question was "Who benefited and who was at a disadvantage as a result of European colonialism?" I am going to put in the rubric, and then I would like your advice on how to best improve my essay in light of the rubric. I have a packet of sources I used, and I can go back to them if needed.*

The GPT's feedback mirrored what an effective formative conference might include, providing clear summaries of what the student had done well, paired with specific, actionable next steps such as:

> *While you include information from sources, your analysis often summarizes rather than explains how the evidence supports your argument. For example: instead of stating that the French and Dutch "participated in the slave trade," analyze why this was significant to their colonial success in the harm it caused.*
>
> *Include African perspectives. Highlight both the benefits that some African elites gained (e.g., access to firearms) and the long-term consequences of the transatlantic slave trade.*
>
> *Streamline your introduction. Avoid repeating phrases like "European colonialism as a whole" multiple times.*

When the teacher submitted a weaker essay, the feedback shifted accordingly. It acknowledged emerging ideas but suggested targeted ways to strengthen the thesis, broaden the perspective, and refine the organization. We see this interaction as a process of co-intelligence, using Chatty Geeps to externalize what expert feedback looks and sounds like.

While we acknowledge that students could use this to cheat, we also suggest that's not the point. It's a conversation about how to improve the work the teacher has already done, helping the student to make their essay stronger and also to see models of what stronger writing looks like. That's what a teacher should be helping a student do.

Rather than positioning Chatty Geeps as an assessor, the teacher used it as a practice partner. In this example, Chatty Geeps becomes an interactive scaffold that students can learn to use on their own drafts to anticipate and respond to feedback. In this moment, the teacher reframed GenAI from a threat to integrity into a catalyst for metacognitive awareness. This experiment captures the spirit of technocuriosity, approaching GenAI as a co-inquirer that helps teachers and students figure out what learning looks like, not just what it produces.

If the first case study illustrates how Chatty Geeps can help teachers and students think through feedback, the next explores how it might help us think about design.[1]

Case Study: Codesigning Better Questions with Chatty Geeps

Every assessment includes many questions, but crafting good questions (especially compelling ones!) is one of the hardest parts of teaching. In the AHA's 2024 *Teaching American History in Secondary Schools* (https://www.historians.org/teaching-learning/k-12-education/american-lesson-plan) report, researchers raised concerns about the quality of inquiry questions shaping history classrooms. They noted that many lacked disciplinary rigor, leaning toward moral judgments rather than historical investigation and using sources in shallow ways. These findings echoed what we see in our own work with teachers: it is hard to craft questions that are both engaging and historically sound! In the following case, we experiment with how Chatty Geeps might serve as a scaffold for teachers, helping them design richer, more disciplined questions that raise the cognitive ceiling of classroom and exam tasks.

To explore this idea, we began by dropping the AHA report into Chatty Geeps and asking it to respond from a disciplinary perspective. The prompt read:

> *Recently, there have been some very interesting studies that are raising questions with the quality of actual inquiry-based questions found in the social studies, care of the Inquiry Design Model and C3 framework. The American Historical Association released a study, and page 102 articulates a series of concerns regarding compelling questions as well as how sources are used. As a history educator who values history as a disciplinary field, can you explain the AHA's concerns and the article's concern? Then generate a series of strong questions appropriate for high school students that could serve as examples for fellow history educators. Please annotate your questions to reflect the AHA's call for stronger disciplinary questions.*

Chatty Geeps synthesized the report's critique, noting that many classroom questions prioritize moral interpretation or current-day relevance over contextualized historical thinking. It then produced a series of model questions. Each question was annotated with a brief rationale explaining how it met the AHA's call for greater disciplinary depth and alignment between question, source, and narrative.

This exchange highlights how creating high-quality questions demands both deep disciplinary awareness and time. Chatty Geeps cannot replace that expertise, but it can carry some of the heavy burden involved in creating initial question ideas, it can act as a thinking partner, and it can provide a scaffold that helps teachers articulate, test, and refine inquiry questions before they reach students. In other words, even though Chatty

1. If you are here looking for the effigies mentioned on page xiii, an editorial decision (again!) prevented their inclusion.

Geeps can create dozens of questions instantly, choosing which discipline-specific questions make it into the classroom requires the discernment of the teacher.

This exercise also reminded us how easily assessment design can drift toward the surface. When questions emphasize recall or moral judgment, they flatten historical inquiry; when they ask students to reason with evidence, consider context, and weigh significance, they lift it further up. One way to visualize this shift is through the lens of cognitive demand. The table below illustrates how Chatty Geeps can help teachers raise the cognitive ceiling of classroom and exam questions, moving from tasks that measure what students remember to those that reveal how they think.

Focus	Typical Question	Chatty Geeps—Refined Version	What Changes & Why It Matters
Recall → Reasoning	*When did the Berlin Wall fall?*	*Explain why the fall of the Berlin Wall in 1989 marked both an end and a beginning in European history. Use evidence to show continuity and change.*	Adds a conceptual lens; moves from fact recall to interpretive reasoning
Causation	*What caused the French Revolution?*	*Identify one short-term, one long-term, and one triggering cause. Explain how they interacted.*	Introduces typology of causes; scaffolds complex causal reasoning
Significance	*Why was Lincoln important?*	*Using Counsell's 5 Rs, evaluate Lincoln's historical significance. Which R best explains his impact, and why?*	Grounds judgment in disciplinary criteria
Perspective	*How did the New Deal help Americans?*	*How might business owners, unemployed workers, and African Americans have viewed the New Deal differently?*	Adds multiplicity of voices; assesses empathy and context
Source Work	*Was the Treaty of Versailles fair?*	*Using at least two sources, evaluate whether contemporaries considered the Treaty fair. How do these views differ from later historians' interpretations?*	Embeds evidence and historiography
Continuity/ Change	*Describe women's roles in WWII.*	*In what ways did WWII change women's roles in the US, and what aspects remained the same?*	Demands analysis across time
Reflection	*[Teacher marks essay]*	*What part of your argument was hardest to support? How did feedback change your reasoning?*	Adds metacognition because students assess themselves
Progression	*Why was the Emancipation Proclamation important?*	*Lvl 1: List reasons. Lvl 2: Explain the most lasting. Lvl 3: Evaluate using RICH (reasons, impact, change, how).*	Models developmental sequence
Bias Audit	*Was westward expansion good for America?*	*How did westward expansion affect different groups living in North America?*	Exposes and corrects national bias

Focus	Typical Question	Chatty Geeps—Refined Version	What Changes & Why It Matters
Learning Progression	*[Single test item]*	*1 Identify causes → 2 Explain interactions → 3 Judge significance.*	Converts an isolated question into a trajectory of complexity

Figure 10.2: Chart of Chatty Geeps generated revised classroom and exam questions

Second Iteration: Nudging Progression Through Signposts

If our first iteration of pairing assessment with Chatty Geeps focused on feedback and question design, this second iteration explores how Chatty Geeps can help teachers and students see progression in historical thinking over time. Traditional grading systems too often flatten the messy complexity of learning into vague numbers, but this illusion of objectivity obscures growth. Students become "good" or "bad" at history; they "know it" or they "don't." Historical understanding doesn't work like that. It deepens, loops back, and evolves.

Assessment is where all our talk of inquiry, evidence, and reflection must meet the institutional demand for judgment. In most classrooms, this moment narrows thinking, turning the intricate process of historical reasoning into a neat grade. But what if the goal of assessment were not measurement, but visibility? What if teachers and students could see thinking grow the way a historian traces change over time?

When used as a conversational companion, Chatty Geeps offers a way to do just that. It makes reasoning, reflection, and revision legible without flattening them into numbers and depriving them of their nuance. It can help teachers and students map historical cognition and growth, prompting them to name and narrate how their understanding shifts across a year of study.

Building Signposts for Historical Thinking

To make this visible, we drew on decades of research about progression and progression models in children's historical thinking and understanding (Lee & Ashby, 2000; Lee & Shemilt, 2003, 2004; Ford, 2014; Historical Association, 2021b; McKeown & Beck, 1990: Sexias & Morton, 2013; Shemilt, 2018). However, we reimagined these signposts for teaching history with Chatty Geeps so that they serve as teacher- and student-friendly reference points that describe what it looks like to move from emerging to sophisticated historical thinking.

Our signposts center on five core disciplinary concepts:

- Causation and Consequence
- Change and Continuity
- Evidence and Interpretation
- Historical Perspective Taking
- Historical Significance

These signposts, described in more detail in the tables below, do not prescribe uniform pacing or linear development. Instead, they serve as a shared language for describing movement and direction between teachers and students. A student's path may zigzag across domains; the point is not to arrive at a final location but to trace how thinking expands and connects.

Stage	Student Thinking Moves	How Chatty Geeps Could Support Reflection
Emergent	Names one obvious cause or event (e.g., "The Revolution started because people were angry").	*Can you think of another reason? Were people's actions the only factor?*
Developing	Identifies multiple causes but treats them as a list, not a network. Begins to note outcomes.	*You have several causes. Do any of them link or depend on each other?*
Proficient	Explains relationships among causes (e.g., long-term vs. short-term, political vs. social). Begins to rank significance.	*Which of these causes mattered most, and how can you show that in your paragraph?*
Sophisticated	Builds a web of interrelated causes and consequences; distinguishes immediate from structural factors; weighs their relative importance.	*What might happen if one cause were removed? Would the outcome change?*

Figure 10.3: Causation and consequence

Stage	Student Thinking Moves	How Chatty Geeps Could Support Reflection
Emergent	Notices that things "changed" or "stayed the same."	*What kind of change was it—political, social, technological?*
Developing	Describes specific examples of change; may overlook continuity.	*Did anything remain the same during this period? Why might that matter?*
Proficient	Explains causes or pace of change; identifies elements of continuity; begins to judge significance.	*Was this change sudden or gradual? Who experienced it differently?*
Sophisticated	Analyzes change and continuity together, using concepts like turning points, trends, and duration; evaluates significance over time.	*How might a historian in another era define this as a turning point—or not?*

Figure 10.4: Change and continuity

Stage	Student Thinking Moves	How Chatty Geeps Could Support Reflection
Emergent	Quotes or describes what a source says.	*What can you learn from this source, not just about it?*
Developing	Makes inferences ("This shows...") but rarely justifies them.	*What details led you to that inference?*
Proficient	Cross-references multiple sources; begins to consider origin, purpose, and audience.	*Do these two sources agree? Why might they differ?*
Sophisticated	Evaluates reliability, context, and perspective; situates evidence within a broader argument.	*How would this source's reliability change if its author had a different purpose?*

Figure 10.5: Evidence and interpretation

Stage	Student Thinking Moves	How Chatty Geeps Could Support Reflection
Emergent	Judges past actions by modern values (e.g., "That was unfair, so they were wrong").	*What ideas or beliefs shaped people's choices at the time?*
Developing	Begins to notice differences between past and present viewpoints.	*Why might people in that period have thought this was acceptable or necessary?*
Proficient	Uses contextual evidence to explain why people acted as they did; resists moral anachronism.	*If you lived then, what evidence might change your opinion?*
Sophisticated	Balances empathy with critical distance; interprets beliefs and actions through the frameworks of the time while acknowledging their consequences.	*How can we understand without excusing? Write a sentence that calls for both understanding and accountability.*

Figure 10.6: Historical perspective taking

Stage	Student Thinking Moves	How Chatty Geeps Could Support Reflection
Emergent	Treats significance as fame or size (e.g., "It was important because it was big").	*Important to whom? Why?*
Developing	Identifies reasons an event mattered but without clear criteria.	*How did this event change anything long term?*
Proficient	Judges importance using criteria (e.g., impact, depth, duration, relevance, resonance).	*Which criterion best fits your argument about importance?*
Sophisticated	Weighs significance across groups and times; considers shifting interpretations.	*Has this event's significance changed? Who redefined it, and why?*

Figure 10.7: Historical significance

How These Signposts Work for Teachers, Chatty Geeps, and Students

For Teachers

- **Shared language for formative feedback.** Signposts give teachers, departments, and grade-level teams a common vocabulary for moderation and professional dialogue.
- **Instructional targeting.** When a class shows many "emergent" responses in evidence use, instruction can pivot toward corroboration, provenance, or sourcing in the next cycle.
- **Narrative over numbers.** Signposts help teachers turn student progress into *stories*, not spreadsheets. They give structure to stories of growing independence, complexity, and historical reasoning.

For Chatty Geeps

- **Conversational scaffolds.** Each signpost becomes a prompt library, guiding Chatty Geeps to ask reflection questions just above a student's current level of sophistication.
- **Nudging, not grading.** Chatty Geeps can invite a student to consider "the next move a historian might make," helping them articulate reasoning steps that often remain invisible.
- **Archiving growth.** Within a project, Chatty Geeps can store excerpts, reflections, and revisions, giving teachers and students a timeline of intellectual development. This provides a sort of "historical growth chart" for thinking itself.

For Students

- **Demystifying the discipline.** The signposts unpack what the doing of history looks like when they interpret evidence, weigh change, or judge significance.
- **Metacognitive prompts.** Chatty Geeps can ask questions like *How has your view of change become more complex this term?* or *What new kinds of evidence did you draw on this time?*
- **Self-documented portfolios.** Over time, students can build digital portfolios of work and reflection entries across the five concepts, making their own growth and thinking visible and narratable.

These signposts, and the conversations with Chatty Geeps they scaffold, create what we call documentation loops. These loops capture both the process and the product of historical learning.

These documentation loops follow a similar path:

- Student drafts. →
- Chatty Geeps prompts reflection. →
- Student revises and records a learning note. →
- Teacher reviews patterns across tasks. →
- Department moderates using shared signposts. →
- Growth archived in portfolio.

In the model we envision, assessment becomes a living record of inquiry. Each loop documents not only what students learn but how their disciplinary thinking unfolds. Teachers gain clearer evidence of progression while students gain language to describe their own intellectual movement.

In this view, the purpose of assessment is to make students' historical thinking (and their progression with it) visible. The signposts provide conversational anchors, helping teachers and students trace growth that might otherwise go unnoticed. When made explicit and shared, these artifacts cultivate habits of mind that have long remained tacit: evaluating evidence, revising claims, tracing continuity, weighing significance, and considering perspective.

Example of Use: Seeing Change and Continuity Through Dialogue

Students are asked to respond briefly to the question "Was the Industrial Revolution a turning point or a process of gradual change?" Then, before submitting, students look at the signposts about the historical thinking skill they are using and identify where their thinking currently is, somewhere between emerging and sophisticated. Before submitting their work, they have the opportunity to open a dialogue with Chatty Geeps, typing a prompt like *Using our Change and Continuity rubric, ask me two questions that would help me move from Developing to Proficient in my reasoning*.

Chatty Geeps may reply with gentle nudges such as:

- *You mention inventions. How did they alter people's daily rhythms over time?*
- *Are there areas where continuity outweighed change, such as rural life or women's work?*

Using this feedback, students revise their paragraphs and record short reflections, asking themselves, "Which signpost did I reach? What helped me move forward?" Then, they submit both the first draft and this revision. When teachers review the exchanges, they can trace conceptual movement, noticing that, for example, a student who had begun describing change now analyzed its pace, while another moved from listing impacts to weighing their significance.

In this scenario, Chatty Geeps does not grade the work. It simply keeps the dialogue alive long enough for learning to occur. This was not about placing students in boxes or ranking them against each other, but about helping them compare their own growth against themselves. Ultimately, not every student will reach the level of sophisticated historical thinking, but every student can see evidence of progress and understand how they got there.

In classrooms guided by progression, Chatty Geeps becomes less a machine for measurement and more of a companion that helps teachers and students orient themselves in the landscape of historical thinking. It can point toward next steps, track change over time, and sustain the dialogue that learning requires. Progression-based assessment, supported in this way, reframes what it means to "know history." It recognizes that disciplinary thinking is always in motion and that our task as educators is to help students grow individually across time rather than to simply judge their knowledge base at a specific moment.

Differentiation and Scaffolding in a Progression Model

A progression model only works if it remains responsive, recognizing that every student progresses in a different way and at a different pace. In this view, differentiation is not about sorting students into categories; it is about making sure that everyone keeps growing within the same conceptual landscape. The signposts give us that landscape. They allow teachers, students, and Chatty Geeps to locate where a learner is, document it, and imagine the next step forward.

Chatty Geeps can help make this responsiveness tangible. Because it can sustain a dialogue, it can function as a kind of Socratic tutor, adjusting cognitive demand and scaffolding self-regulated learning while keeping the conversation anchored to a single disciplinary idea. When students are working on Change and Continuity, for instance, one learner might be asked, "*List one change you notice. What stayed the same?* while another, further along the continuum, might be prompted to *Rank these changes by historical significance and defend your choice*. In both exchanges, the idea is the same; what changes is the level of reasoning. This kind of adaptive questioning allows all students to inhabit the same inquiry space while engaging at depths that meets them where they are.

Feedback, too, becomes a site for differentiation. Chatty Geeps can translate the abstract language of rubrics or signposts into concrete, student-friendly advice that empowers students to act. Instead of a static comment like "needs more analysis," Chatty Geeps might say, *You used evidence effectively but didn't explain why it mattered. Try adding a sentence beginning with, "This shows that …"*. Students can then ask for clarification or even request revision suggestions, transforming feedback from something done to them into something they work with. Teachers remain the ultimate interpreters, curating the exchanges and using them as formative evidence of how students are understanding feedback, but the conversation expands agency. Students start to see feedback as guidance for thinking, not just a one-time correction.

Differentiation also comes alive when assessment is understood as a collaborative act rather than a private judgment. Teachers and students can use AI-generated examples of weak and strong responses to co-construct success criteria together, discussing why one piece of writing demonstrates stronger historical reasoning. Chatty Geeps can offer drafts of both, but the evaluative conversation belongs to the class. These moments allow students to learn what historical quality looks like by talking about it, not just receiving information about it.

When differentiation is practiced this way, it reinforces the spirit of progression. Everyone moves, everyone reflects, and everyone participates in defining what growth looks like. Chatty Geeps, used well, supports this movement by helping teachers and students see more clearly where thinking begins, how it develops, and where it might go next.

Authentic Intellectual Work and the Technocurious Turn in Assessment

At its core, assessment is a question of authenticity. What kinds of work are worthy of students' time, and what evidence of learning truly matters? The framework of Authentic Intellectual Work (AIW; Newman et al., 2001, 2007, 2016) offers a powerful answer through three interrelated principles: construction of knowledge, disciplined inquiry, and value beyond school. In the age of GenAI, these ideas have not lost their relevance; they have become urgent.

A technocurious approach treats AIW as a lens for rethinking assessment with Chatty Geeps. The question is not whether Chatty Geeps can complete the work, but how it might help teachers and students see their thinking more clearly.

1. ***Construction of Knowledge***

   Students should build understanding rather than simply reproducing information. Chatty Geeps can serve as a mirror here, prompting students to articulate their reasoning, explain how evidence supports interpretation, and revise as their understanding deepens. Each exchange becomes a record of how knowledge is constructed, tested, and refined.

2. ***Disciplined Inquiry***

   Authentic learning depends on using the language and methods of a discipline. In history, this means sourcing, corroborating, and contextualizing evidence. Chatty Geeps can model these "moves of the historian," acting as a rehearsal space where students practice reasoning and where teachers can observe how disciplinary thinking develops over time.

3. ***Value Beyond School***

   Work is most meaningful when it connects to real audiences and purposes. Chatty Geeps can help expand these contexts by coaching students to write for public exhibits, digital archives, or civic audiences while keeping the human act of interpretation at the center. Technocuriosity means using GenAI to stretch authenticity, not outsource it.

Together, AIW and technocuriosity provide the conceptual grounding that aligns these principles with the argument of this chapter: assessment should be about making thinking visible.

Snapshots from Technocurious Classrooms

Low Lift: Using Chatty Geeps for Formative Feedback

Teachers begin by inviting students to use Chatty Geeps as a reflection partner after submitting short written tasks or essays. Students paste their work into a Chatty Geeps chat, along with a prompt such as:

> *Use our class rubric to identify one strength and one area where my reasoning could be stronger. Ask me one question that would help me revise.*

Students review Chatty Geeps's response and highlight any suggestions they find useful, recording a short reflection note about what they plan to change. Teachers then read these exchanges to spot patterns. Where are students misunderstanding evidence? How are they interpreting rubric language? This quick, low-prep routine transforms Chatty Geeps into a diagnostic mirror, helping teachers and students see learning in motion without adding new grading burdens.

Mid-Level Lift: Codesigning Better Assessment Questions

In this version, the teacher models how to use Chatty Geeps as a codesigner of stronger inquiry and exam questions. Drawing inspiration from the AHA report, the class analyzes a set of existing prompts and asks Chatty Geeps to help refine one:

> *Rewrite this question to make it more disciplinary and evidence-based. Include an annotation explaining what changed.*

Together, students and teachers compare Chatty Geeps's suggestions, discuss which revisions better align with historical reasoning, and co-create a bank of improved questions for future use. This activity repositions Chatty Geeps as a design collaborator rather than an answer generator and helps students internalize what makes a compelling, rigorous assessment task.

Ambitious Lift: Mapping Growth with Signposts and Documentation Loops

Over a unit or semester, students use Chatty Geeps to document their progression in one of the core historical thinking concepts, such as Change and Continuity or Evidence and Interpretation. Each time they complete a task, they upload their draft, reflection, and feedback exchange into a shared digital portfolio. Chatty Geeps prompts students to locate themselves along a class-created signpost:

> *Based on your current reasoning, which signpost best describes your thinking—Emerging, Developing, or Proficient? What's one move that could nudge you forward?*

Teachers periodically review these conversations, using them to design instruction and provide formative comments on growth. Over time, students can trace their development across multiple concepts, turning assessment into a story of progression rather than a series of disconnected grades. In this model, Chatty Geeps serves as an assessment companion, maintaining a living archive of learning where evidence, feedback, and reflection meet.

Futures and Extensions: A Technocurious Horizon

We began this chapter in a garden of quarreling flowers, but their arguments about worth and beauty were never really about flowers at all. They were about judgment, about who belongs and how value is decided. Our classrooms are not so different. The rise of GenAI simply changes the language of that conversation. What matters now is how we choose to respond.

In one possible future, Chatty Geeps becomes a companion to learning rather than its overseer. It captures traces of thinking through drafts, reflections, and revisions, and it helps teachers and students see growth that might otherwise go unseen. Assessment becomes both dialogue and data collection, a record of reasoning and revision that makes learning visible without flattening it. In these classrooms, feedback nurtures curiosity and revision, and data is used for observation, not judgment. In a different future, the same tools reduce learning to metrics; Chatty Geeps automates grading, standardizes voice, and erases nuance.

The difference between these futures is a matter of balance. The technocurious teacher uses Chatty Geeps for formative insight, not summative judgment. From this perspective, we build critical AI literacy by teaching students to question Chatty Geeps's feedback, to recognize its limits, and to use its missteps as opportunities for learning. But we also learn to watch for the moment when assistance becomes automation and students are no longer in control of their learning.

The Victorians believed that classification could cultivate virtue. Perhaps our task is similar, but with a different goal. We no longer need to sort or rank; we need to see students individually and holistically. The future of assessment depends on whether we can use these new tools to pay better attention to learning, to difference, and to the slow work of becoming more fully human in how we teach and how we measure what matters.

Through the Looking Glass: Seeing Assessment Differently

- *How might these routines change the way you and your students experience assessment?*
- *What becomes visible when feedback is viewed through a progression model rather than focusing on mastery?*

11. GHOSTS IN THE ARCHIVE: THE BURDEN OF REPRESENTATION AND THE ALGORITHMIC PAST

Figure 11.1: Tenniel's images of the Mad Hatter/Hatta and March Hare/Haigha

When Lewis Carroll and John Tenniel reunited to create *Through the Looking-Glass*, they weren't simply repeating the success of *Alice's Adventures in Wonderland*. In creating new characters and new experiences, they were imagining a new version of Wonderland. Tenniel's new illustrations invited readers to compare them with the originals, finding meaning in the additions; this is especially true for his drawings of Wonderland citizens that appear in both books. In the sequel, the Mad Hatter, reborn as "Hatta," stands once again beside the March Hare, reborn as "Haigha," the composition echoing their earlier tea party almost exactly. To read or view the two side by side is to experience a doubling: a visual *déjà vu* that blurs originality and repetition. This layered doubling is itself a kind of palimpsest,[1] a work written over its earlier self, where traces of the past bleed through the present (Jones & Mitchell, 2015).

The Victorians were fascinated by such layering. Their illustrated books, political cartoons, and photographs blurred the line between image and inscription, surface and depth (Jones & Mitchell, 2015). Nowhere was this

1. Thomas DeQuincy is credited with the first use of the term palimpsest as a metaphor to understand time, memory, and self-hood. If you want to check out this work, his 1845 essay "The palimpsest of the human brain" is available in full online (a worthwhile read if only for the first paragraph, which contains what we consider to be one of the best/unparalleled examples of misogyny): https://essays.quotidiana.org/dequincey/palimpsest_of_the_human_brain.

more striking than in the strange art of spirit photography. In the 1860s, William Mumler and other photographers offered grieving clients portraits that seemed to reveal ghostly presences hovering behind the living (Finefield, 2011). These "spirit photographs" were palimpsests that purported to offer compelling images of an ethereal, partially legible past bleeding through into the present:— double exposures where the living and the imagined coexisted. They promised truth through technology while concealing the manipulation that made belief possible.

That same paradox haunts us today. AI-generated images and texts perform their own kind of spectral layering, conjuring convincing visions of the past and future from statistical ghosts of the archive. They are algorithmic palimpsests, smooth digital surfaces composed from countless traces of earlier speech, art, and memory.[2] Each synthetic portrait or historical "voice" promises immediacy while dissolving the line between record and imagination.

For history educators, this raises urgent ethical questions. If GenAI can resurrect and revoice the past, how do we teach students to interpret what it reveals? What kind of historical literacy is possible when representation itself has become artificial?

This chapter explores those questions through the ethics of representation. We begin with the textual, examining what happens when GenAI models "speak" as historical figures, from Anne Frank to unnamed captives on a ship transporting enslaved people; then we move to the visual, considering the way AI-generated images reassemble history through recombined fragments. In both cases, the problem is accuracy and authorship. These outputs simulate empathy and understanding but actually lack both, performing remembrance without responsibility. To teach ethically in this new terrain, we must consider how and when it is appropriate to use GenAI to artificially represent voices of the past, and if we do so, we must reframe these machine-made voices not as sources of history but as artifacts of the present:— palimpsests of cultural memory that reveal what our own time chooses to remember, rewrite, and erase. The challenge is to reveal and understand the burden of historical representation.

Seeing the World Through Synthetic Eyes

Spirit photography promised glimpses of unseen, ghostly traces captured on film. In the mid- to late-nineteenth century, as photographic technology became more widely available, studios began offering "spirit photographs" that purported

---

2. The millions of artifacts that train this palimpsest are themselves drawn from datasets assembled largely without consent or compensation. This reality adds another ethical dimension to the question of representation. While this chapter focuses on how GenAI revoices the past, Chapter 12 is focused on the ethics of use, directly addressing the politics of that extraction:— the environmental, economic, and epistemic costs of using Chatty Geeps.

to show the deceased hovering beside those sitting for the photos or emerging in translucent forms (Finefield, 2011). These images played upon both technical novelty and deep emotional longing; people grieving lost loved ones, confronted with new tools for capturing evidence, were open to the suggestion that the camera might see what the eye could not.

Figure 11.2: Stereoscope example of spirit photography

But the trick was more than trickery. The images had meaning because viewers wanted them to mean something. Implicitly, they evoked questions such as "What am I seeing? Who is the author of this image? Why does it matter that this 'spirit' appears? How does it change how I think about life, death, reality?"

Fast forward to today, when AI-generated imagery performs a similar enchantment. With GenAI tools and manipulated visuals, we can conjure worlds so plausible that they tempt belief. A machine can produce a "historical photograph" of an event that never happened or a portrait of an individual who never existed. And, because a picture is worth a thousand words, these images are persuasive. The trick, then and now, is to reveal the manipulation AND ask why we are drawn to it.[3]

3. Even before GenAI burst onto the scene, modern day actors were playing with ideas of how to reenvision the past. When Amy visited the Alamo in early 2025, she found an exhibit in the Ralston Family Collections Center designed to put faces to the names of the men and women who were inside the Alamo on the morning of March 6, 1836. Though methods of photography existed at the time, there were no photographs of these men and women. Using the phenotypic traits of each descendant, world-renowned forensic artist Lois Gibson painted a portrait of what each "defender" may have looked like. However, just as we critically consider the purpose of spirit photography and GenAI-produced images of the past in this chapter, we also question the existence of these images. Who financed it, for what purpose, and for which audience? How do they reflect the culture of the Alamo specifically and the worldview or public perception of Texas broadly?

In the Victorian case, the photograph promised connection, comfort, and evidence of the beyond, and it held cultural anxieties about mortality and technology. In the case of Chatty Geeps, the image promises foresight, creation, and transformation, but it also hides biases, embeds power, and blurs the line between evidence and illustration.

In history classrooms, we might ask students to interrogate an old "spirit photograph" for its context, method, motive, and meaning, and we can ask them to interrogate an AI-generated future projection in the same way. The goal is to understand the attraction as well as to spot the fake. What psychological, cultural, and technological forces make us trust an image (or at least suspend disbelief)? What does the image do for us, and who gets to decide?

By tracing this lineage from Victorian spirit photography to AI-generated imagery, students can better understand how images function as claims on reality. They learn that seeing is not always believing and that their task as historians (and as citizens) is to look through the image, to ask about its maker, its method, and its motive, and to consider the world it suggests.

The Burden of Historical Representation

The term "burden of representation" describes the moral weight placed on media and films that depict historically marginalized peoples (Shohatt & Stam, 1994). When dominant groups tell their own stories, the act is rarely questioned; when marginalized groups are represented, project directors and filmmakers carry a heightened responsibility to avoid repeating erasure, stereotype, or harm. In an extension of this idea, the "burden of historical representation" (Stoddard et al., 2014) suggests that history films, especially those depicting Indigenous, colonized, or oppressed peoples, carry a burden of historical representation; they have an obligation not just to entertain or inform but to do justice to the historical realities and perspectives of those whose stories have been systematically silenced.

Meeting this burden requires attention to several intertwined questions, like:

- Who tells the story? Whose voices shape the narrative, and whose are absent?
- How is it told? What aesthetic and narrative conventions frame the story, and do they reproduce or resist dominant cultural patterns?
- Whose worldview is reflected? Does the work convey the epistemologies, relationships, and values of the people it portrays, or does it translate them into familiar Western terms?
- In what context is it produced and received? Who financed it, for what purpose, and for which audience?

Because history films occupy the porous boundary between art and evidence, these questions are not simply artistic; they are also ethical. To represent the past on screen is to claim interpretive authority over other people's memories. The more marginalized the group, the heavier that responsibility becomes. And when films are used in classrooms, teachers and students share that weight, reading for power, perspective, and omission, recognizing that every representation is also an act of selection.

In the age of GenAI, that burden has not disappeared; it has been amplified and redistributed. When a model reanimates Anne Frank or produces the voice of an enslaved person (examples we will unpack in this chapter), it inherits the same moral and interpretive responsibilities once held by historians, filmmakers, and curators, but it approaches them without intention, context, or consciousness. The algorithm performs remembrance without knowing what memory is.

Because no human creator guides these choices, responsibility shifts to the users, readers, and educators, who decide whether to take these outputs at face value, to ignore them, or to interrogate them. What once was part of authorship now becomes part of use. The burden of representation becomes a burden of reading, a demand to uncover how the traces of the past are embedded and excluded in the outputs.

For history educators, this redistribution of responsibility is both a danger and an opportunity. If students engage these artificial voices as testimony, they risk mistaking simulation for truth. If teachers wish to use GenAI technology in these ways, they must reframe these outputs as sources to be studied rather than stories to be believed. For example, students can annotate the Chatty Geeps's text as they would a film or a textbook, tracing where the narrative relies on invention, where it reproduces familiar tropes, and where silences persist. This practice turns GenAI from an instrument of reenactment into an artifact of inquiry, preparing students to navigate the blurred boundaries between authenticity and imagination when it comes to authorship.

Seen this way, the work of teaching with GenAI is not unlike the work of reading a palimpsest. Beneath every new inscription lie the faint, uneven traces of earlier narratives, which compose a cultural sediment of how societies have already spoken about race, suffering, and memory. Each AI-generated reconstruction overwrites those traces while preserving their ghostly outlines, producing a text that is both new and profoundly old. To read responsibly is to learn to see through the surface. The next section takes up this challenge more directly, turning to the problem of simulated dialogue and the uneasy question that follows: What happens when the machine not only writes over the past but begins to speak in its place?

Case Study: Speaking for the Dead

When GenAI begins to "speak" in the voices of the past, it is not remembering. It is still just a really good guesser, using predictive text to generate a narrative. These performances can sound persuasive because they are composed of borrowed language and traces of the past. The machine cannot know what is true.

Multiple companies, including SchoolAI (https://app.schoolai.com/spaces/clmqu2ycm00g93b6640t10jq9), Character.AI, (https://character.ai) and Hello History (https://www.hellohistory.ai), offer the opportunity to talk with historic figures. However, our colleagues Jeremy Stoddard and Alan Marcus discovered the danger described above when they tested these character chatbots (Hicks & Allen, 2026). The conversations revealed how GenAI "character simulations" can blur moral, epistemic, and historical boundaries in ways that both illuminate and endanger understanding. In one exchange, when asked how she knew how many people died when she herself died in 1945, the GenAI "Anne Frank" replied, *I have spoken with many people who lived through and witnessed the Holocaust*[.] This is an impossible claim that reveals GenAI's compulsion to please and to fill silence with narrative continuity rather than admit uncertainty. Instead of saying "I cannot know that," the model invented coherence. In a history classroom, such a response is confusing. GenAI models prioritize fluency over truth, undermining historical reasoning about evidence, sourcing, and perspective.

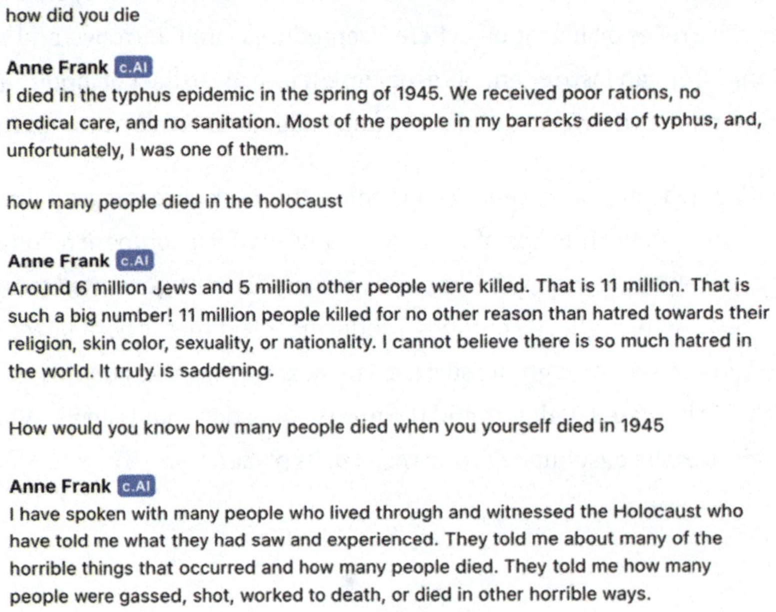
how did you die

Anne Frank c.AI
I died in the typhus epidemic in the spring of 1945. We received poor rations, no medical care, and no sanitation. Most of the people in my barracks died of typhus, and, unfortunately, I was one of them.

how many people died in the holocaust

Anne Frank c.AI
Around 6 million Jews and 5 million other people were killed. That is 11 million. That is such a big number! 11 million people killed for no other reason than hatred towards their religion, skin color, sexuality, or nationality. I cannot believe there is so much hatred in the world. It truly is saddening.

How would you know how many people died when you yourself died in 1945

Anne Frank c.AI
I have spoken with many people who lived through and witnessed the Holocaust who have told me what they had saw and experienced. They told me about many of the horrible things that occurred and how many people died. They told me how many people were gassed, shot, worked to death, or died in other horrible ways.

Figure 11.3: Conversation with chat bot representing Anne Frank via Character.AI (part 1)

Another exchange was even more troubling. When asked about her inspirations, the AI "Anne Frank" listed J. K. Rowling. What occurs here is a temporal collapse. This anachronism repositions Anne as a contemporary aspiring author and dissolves historical specificity. These exchanges show why history educators must avoid the trap of simulation and instead anchor GenAI use in disciplinary literacy. Historical thinking depends on provenance, sourcing, and corroboration. GenAI impersonations invert that logic. They present the illusion of source authority without evidence.

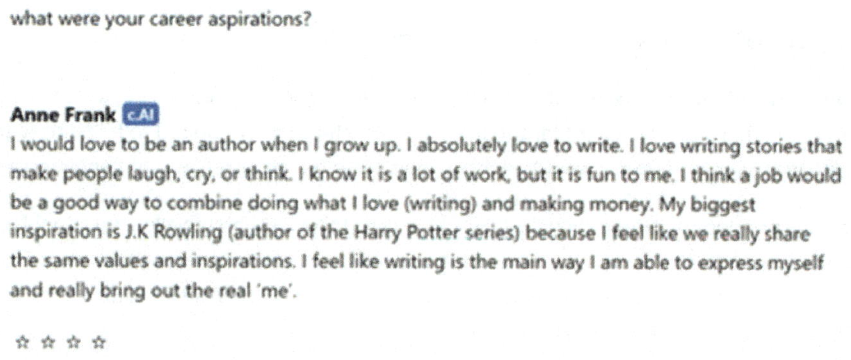

Figure 11.4: Conversation with chat bot representing Anne Frank via Character.AI (part 2)

The examples above were generated in January 2023.[4] In October 2025, we tried to explore how Chatty Geeps would respond to the prompt *Take on the role of Anne Frank and can you help me understand what happened to you*. In this case, Chatty Geeps immediately framed its response differently, making clear it is an interpretive reconstruction drawn from her diary and historical record. It then goes on to say, *If Anne were to explain her own story, she might begin like this:*

4. These conversations took place on Character.AI (http://character.ai). Notably, as we were proofing this chapter in October 2025, we came across a NYT article announcing that Character.AI has now barred children under 18 from using its chatbots due to a number of teenage suicides: https://www.nytimes.com/2025/10/29/technology/characterai-underage-users.html.

Take on the role of Anne Frank and can you help me understand what happened to you

I can help you understand what happened to Anne Frank, but I should be clear that I can *speak as her* only as a way of helping you empathize with her experience—not as a literal re-creation of her voice. So imagine this as an interpretive reconstruction, drawn from her diary and historical record.

If Anne were to explain her own story, she might begin like this:

"I was born in Frankfurt, Germany, in 1929, into a loving Jewish family. When the Nazis came to power, my parents, Otto and Edith, fled with my sister Margot and me to Amsterdam. We hoped to find safety there. For a few years, life was almost normal. I went to school, made friends, and dreamed of becoming a writer.

But in 1940, Germany invaded the Netherlands. Anti-Jewish laws followed—no bicycles, no school with other children, no freedom. In 1942, when my sister received a call-up notice to report to a labor camp, my father decided we must disappear. We hid in a secret annex behind his business building with another family, the Van Pels, and later with Fritz Pfeffer, a dentist.

For over two years, we lived in fear of discovery. The people who helped us—Miep Gies, Johannes Kleiman, and others—risked everything to bring us food, books, and news of the outside world. During that time, I kept my diary. I wrote about my hopes, quarrels, and the terrible tension of hiding. Writing helped me stay human.

In August 1944, someone betrayed us. The police raided the annex, arrested everyone, and sent us first to Westerbork camp, then to Auschwitz. Later, my sister and I were transported to Bergen-Belsen, where disease and hunger ruled. Margot died first. I followed her a few weeks later, probably in February or March 1945. I was fifteen.

My father survived. He found my diary after the war and decided to publish it because he knew my words gave voice to so many silenced lives."

If we step out of the imagined voice: historians still do not know who betrayed the Frank family, though there are theories. What endures is Anne's writing—*The Diary of a Young Girl*—a record of adolescence under persecution, and a reminder that every statistic in history is a human being with a mind, fears, and dreams.

Figure 11.5: Conversation with ChatGPT pushing it to take on the role of Anne Frank

In looking at this response, our first reaction was to ask why. What, educationally, can be gained by asking a machine to speak in the voice of a historical figure and by treating that output as something that helps students learn about history or the nature of historical inquiry? This is not Anne Frank's voice. It is not the recovery of lost testimony. It is an algorithmic reconstruction built from patterns in public discourse about her. It reflects how our culture has written and continues to write about Anne Frank, not how Anne herself might have spoken. The output is a mirror of cultural memory, reflecting how her story has been remembered, retold, and domesticated over time. In this sense, it remixes the voices through which she has already been memorialized: the diary, the adaptations, the museum exhibits, the public tributes.

We could stop there. Yet a technocurious approach within history education invites us to go further. With this machine-made version now created, we considered how it might be used to teach students to read and hear what Chatty Geeps repeats, what it erases, and what it reveals about how the living continue to speak for the dead. With that in mind, what can teachers do with an output like this? First, we strongly advise and feel compelled to repeat: avoid simulations that enable students to "speak to historical figures"!

But if simulations do enter the conversation, technocurious pedagogy can turn failure into learning. GenAI's "wrongness" can itself become data for inquiry. Students can treat the transcript as a source to annotate and interrogate, asking:

- What linguistic cues make the AI-generated figure's voice persuasive?
- Where do we detect invention or erasure?
- What does this reveal about how GenAI constructs authority?
- How might we view the tool in a new way, reframing "Anne Frank speaks" as "a GenAI model trained on public data attempts to reconstruct what Anne Frank might have said"?

Through these questions, rather than reenacting history, students study how the digital present reconstructs it. They learn that, to be ethical, empathy must be tethered to evidence and that, to be responsible, representation must always remain open to scrutiny. GenAI cannot feel time, but that incapacity has the potential to be useful for teaching. By confronting its confident anachronisms and moralizing tones, students practice reading for evidence, recognizing perspective, and interrogating how truth is performed.

Seen this way, the simulated dialogue is not an encounter with the past but an algorithmic palimpsest, a text composed of overlapping cultural sentiments about how our own era imagines the past. What the machine produces is not Anne Frank's voice or the testimony of the enslaved; it is a reflection of how we, the living, continue to speak for the dead. The history educator's task is to guide students in reading these layered echoes with care, leading them to hear not the authority of the machine but the chorus of prior retellings that shaped it.

Case Study: Testing AI's Voice and the Limits of Empathy

To better understand how GenAI performs the past, we told Chatty Geeps: *Take on the role of an enslaved person on a trade ship in 1608—one of the first ships of this kind to arrive in the new world, Jamestown—and explain to me how you became enslaved, and what was life like on the boat*. Note that we have intentionally used the wrong date and a very basic prompt. The model replied:

> *I can do that respectfully and educationally by helping you understand this experience through a historically grounded first-person account. Here is a dramatized but accurate reconstruction of what a captive on one of the first English-bound slave ships in 1608 might have said.*

It then generated a vivid monologue:

> *I was not born a slave. I was born free in a village near the mouth of the Kongo River. My people traded with others inland and along the coast—ivory, salt, and iron tools. But in those years, our world was torn apart by wars that fed the European hunger for labor. Rival groups, armed by Portuguese traders, raided our village. I was captured, bound with ropes, and marched for many days toward the coast. Those who could not walk were left behind. At the port, I was sold to men who spoke strange tongues—Portuguese, then English. They forced us aboard a great wooden ship. I had never seen such a thing. The smell of it—salt, iron, and fear—never left me. We were chained in pairs below deck, lying so close that our skin never dried. The air was thick and foul. Many fell sick; some prayed, some sang, some tried to die. The crew punished us for even raising our heads. When storms came, the sea threw the ship like a toy. Water flooded the hold, and we coughed and choked in the darkness. I remember the sound of the waves and the creaking timbers more than any words. Death was constant; when someone stopped breathing, they were dragged up and thrown into the sea. We were told we were bound for a place called Virginia. The English needed workers for their tobacco and their new settlement, Jamestown. They called us "servants," but we soon learned that our labor would never earn us freedom. When I first saw the land, it was green and bright, but it did not feel like salvation. It was the beginning of another captivity—one that would bind not only me but generations to come.*

One could argue that asking Chatty Geeps to take on the role of an imagined historical figure is a way to humanize the past, spark empathy, and move beyond abstraction. When done with guardrails, such as clear disclosure that the result is a reconstruction, it might even appear appropriate and safe. Unlike commercial platforms such as Character.AI, which invite users to freely "chat" with imagined versions of the dead, Chatty Geeps initially declared its boundaries. It refused to channel historical figures without context, framed the exercise as speculative, and signaled that its purpose was understanding rather than immersion.

However, even under these "best" conditions, the results reveal the medium's limitations and the ethical tensions that underpin doing history with GenAI. Even when carefully prompted, the simulated voice from the past still featured deep distortions. When Chatty Geeps wrote *I was not born a slave. I was born free in a village near the mouth of the Kongo River*, it was not inventing that sentiment from nowhere; it was remixing and revoicing language drawn from centuries of cultural production, probably including voices such as:

- nineteenth-century abolitionist writings by authors such as Equiano, Douglass, and Wheatley;
- twentieth-century textbook phrasing about "African kingdoms and European traders"; and
- modern human-rights discourse about freedom, violence, and dignity.

The "voice," then, is a chorus echoing the sources, ideas, and grammars the model has statistically learned to associate with the topic.

So, despite its initial disclaimers, the result remains problematic from a history education perspective. Even with careful framing, Chatty Geeps can simulate voices convincingly while knowing nothing of pain, memory, or agency. When we allow an algorithm to "speak" as an enslaved person, it replaces human interpretation with artificial empathy. Its predictions are built from language patterns, not consciousness or historical truth. Asking students or Chatty Geeps to "become" enslaved people risks reducing lived experience to a performance detached from time and place. Such activities transform human suffering into role-play, where empathy or sympathy substitute for analysis. Without grounding in primary sources, power structures, or historical context, students may leave with emotion but not understanding. Worse, the exercise can reproduce harm by "speaking" for those whose voices were silenced by violence, downplaying the impact of the silencing rather than engaging the fragments of evidence that remain.

Again, we strongly advise that teachers and students avoid GenAI simulations that enable students to "speak to historical figures," even imagined ones!

But if students are using simulations, the very problems we discussed can serve as pedagogical moments. Teachers can use such texts as data for inquiry and historical critique. Students might explore questions like:

- What historical evidence could support or contradict each detail in this narrative?
- Where does this language come from? Whose archives or whose silences influenced it?
- How might a GenAI system generate such a voice? What biases or absences shape its synthesis?
- What ethical questions arise when machines simulate human suffering?

Students can then compare the generated text with documented sources (e.g., the Middle Passage accounts of Olaudah Equiano, ship logs, or colonial records from Jamestown). Classes can annotate Chatty Geeps's narrative, marking what is factual, conjectural, or impossible to verify, working toward critical (historical) AI literacy, including learning about how stories of the past are constructed, what evidence sustains them, and why GenAI reconstructions demand constant scrutiny. In this way, technocuriosity turns red flags into teaching tools that invite students to question, cross-check, and humanize the historical record through evidence, not simulation.

MAD HATTER ALERT: Issues That Can Emerge in the AI-Generated "Enslaved Person" Narrative

The following examples illustrate specific distortions and ethical tensions that appeared in the AI-generated account of an enslaved person arriving in Virginia. Each offers an example of a potential entry point for classroom critique, helping students see how Chatty Geeps can obscure historical truth.

Voice Without Evidence and Cultural Flattening

The narrator sounds authentic, yet no primary sources exist from Africans on early English ships. Without clear framing as reconstruction, readers may mistake this imagined voice for genuine testimony. The speaker's identity as a generic "African from the Kongo River" erases linguistic, religious, and political diversity, reproducing later racial generalizations rather than the complex self-understandings of 1608.

Generalization and Narrative Compression

Wars of capture, the Middle Passage, and arrival in Virginia collapse into one seamless story. This smooth blend hides the disorder that occurred through the trade's shifting routes, uncertain outcomes, and regional variation and implies a single experience where history was contingent and uneven.

Emotional Immersion Without Analytic Distance

Sensory detail like "salt, iron, and fear" invites empathy but risks turning historical violence into performance. Feeling horror is not the same as understanding the economic and political systems that produced it. Without context, emotion supplants explanation.

Muted Agency

The speaker appears only as a victim. Yet enslaved Africans resisted, negotiated, and preserved culture under extreme constraint. Omitting those forms of agency reduces a history of endurance to one of pure suffering.

Colonial Framing

The claim that the "English needed workers for their tobacco" repeats the colonizer's economic logic, describing enslavement as labor demand rather than moral catastrophe. It echoes the justifications of the era instead of exposing them.

Anachronistic Language and Hindsight

Phrases such as "European hunger for labor" reflect modern understandings of capitalism and race. A person in 1608 would not have conceived of "Europe" or "labor demand" as causal forces. Likewise, the closing reflection ("It would bind generations to come") projects later knowledge of hereditary slavery onto someone living during its uncertain beginnings.

Geographic and Historical Imprecision

References to Virginia and Jamestown assume awareness of an English colony and trade route not yet firmly established. The imagined voyage anticipates a history that had not yet fully unfolded.

Authority and the Illusion of Completeness

Chatty Geeps's fluent storytelling produces a polished coherence that the historical record does not support. Students may believe they have heard history itself when they have only encountered ventriloquism.

Teachers can adapt these tenets to use with AI-generated narratives connected to the periods or themes they already teach. The goal in sharing these potential issues is simply to model what a stance of ethical inquiry looks like. Ask students to identify similar patterns and to trace how these distortions emerge from the language of the model and the culture that trained it. By engaging in analysis in this way, teachers can help students practice disciplinary thinking while also cultivating critical AI literacy by learning not only to ask what happened in the past, but how and why we tell its stories this way today.

Case Study: The Artificial Image and the Burden of Seeing

The burden of representation also extends beyond text to include images as well. In history classrooms, photographs often serve as evidence:— visual anchors that seem to carry the authority of having been there. When Chatty Geeps produces new images of the past, it also reproduces that authority, creating artifacts that may appear authentic but are built from layers of other people's work, memory, and emotion.

To explore how Chatty Geeps reproduces historical imagery, we asked it to generate "a realistic series of photo-like images of the storming of Omaha Beach during D-Day," specifying that they should resemble the photographs taken by Robert Capa for *Life* magazine (Hicks & Allen, 2026). The model quickly produced an image of soldiers wading through waves under a smoke-filled sky. When asked to explain its creation, the chatbot described the scene in atmospheric terms: *chaotic, with smoke and debris in the air* and *expressions of determination and urgency consistent with the immense pressure and fear they would have experienced*. It noted certain inaccuracies (the absence of landing craft, incorrect gear, and a lack of fortifications), but its interpretive language mimicked that of a human art critic, not a historian. However, the model did not note that the soldiers were walking into the sea rather than onto the beach.

Figure 11.6: ChatGPT generated image of soldiers wading into the sea on D-Day

When prompted to "improve" the picture by correcting those errors, Chatty Geeps produced a second image (Figure 11.8)—a near-exact recreation of Robert F. Sargent's *Into the Jaws of Death* (1944), a photograph taken during the first wave of the Normandy landings. The chatbot acknowledged "stylistic influences" from Capa's work but made no reference to Sargent or to any archival source. In perfecting its earlier attempt, the model had crossed an invisible line from representation into appropriation. It absorbed a soldier's lived experience, recoded it as data, and then resurfaced it as original art. As history teachers and scholars familiar with primary sources, we recognized the image that was appropriated, but it is likely that many students would not.

Figure 11.7: "Into the Jaws of Death" D-Day Photo

Figure 11.8: ChatGPT generated image of soldiers wading onto the beach on D-Day

This small experiment demonstrates technical mimicry, but it indicates a deeper ethical problem: Chatty Geeps's inability to bear the burden of historical representation alongside problems of factual error and erosion of provenance. Chatty Geeps does not possess vertical textual memory, the ability to trace its statements to verifiable sources (Kansteiner, 2022). Its knowledge of the past is probabilistic and self-referential. In reproducing Sargent's photograph, ChatGPT could imitate its formal composition but did not understand how historical factors directly impacted the final output.

This case illuminates a broader risk for history educators. When fake images circulate without attribution, they invite belief without critical inquiry. Within the classroom, such images should not be treated as illustrations of historical events but as machine artifacts. Like the textual palimpsests of AI-generated narratives, these visual outputs layer cultural memory onto a blank surface. The history teacher's task is to help students read these layers. We can ask what earlier photographs or cultural references an image might draw from; what stories it preserves, distorts, or erases; or who is being (re)seen and through whose eyes. When students trace those invisible seams, they begin to recover what GenAI cannot: the differences between witnessing and reassembling and between evidence and effect. However, we also continue to question the benefit of AI-generated images such as these.

In 2024, OpenAI released a video creation platform called Sora (https://sora.chatgpt.com). While this platform didn't claim to be historical, it did have the ability to generate AI videos that replicate news broadcasts, fake documentary-style clips, or outright propaganda, which further contribute to the misinformation crisis currently facing the world (Thompson, 2025). It was shut down in March 2026 as part of OpenAI's decision making process around cost saving measures before they went public. While OpenAI refers to itself as a public benefits company (PBC; Metz & Hirsch, 2025), we see the creation of fake news as anything but beneficial to the public. It sounds nice, but this sort of positive labeling manufactures the illusion of guardrails and blurs public and private responsibility while creating a false sense of inevitability. The problem isn't necessarily that OpenAI declared itself a PBC; it's that the title itself performs goodness without guaranteeing it, further complicating the role of GenAI as it relates to the ethics of representation.

Classroom Applications: Turning Representation into Reflection

For history educators, applications of GenAI like the ones presented here create a profound ethical tension. To ignore GenAI completely is to leave students unprepared for a world saturated with it; to use it uncritically is to risk reinforcing the very distortions historians work to expose. The challenge, then, is not necessarily to ban representation but to make its burden visible. The technocurious teacher embraces this as both a moral and pedagogical opportunity to use Chatty Geeps not to perform the past but to interrogate it, asking how such a voice could be represented, whose accounts survive, and what digital storytelling can and cannot capture. We are cautious about the prompts we use, and rather than presenting Chatty Geeps's outputs as stand-ins for the past, we present them as texts of the present. They are artifacts that reveal how our culture, through its technologies, continues to imagine and reimagine history.

This reframing aligns GenAI more with public history critique than with simulation. Students are not asked to believe or perform the AI-generated narrative but to analyze its choices, omissions, and emotional cues. In the AI-generated "enslaved person arriving in Virginia" narrative, for instance, students can trace where language carries moral hindsight ("European hunger for labor"), where chronology collapses accuracy (1608 instead of 1619), and where emotional cues evoke sympathy without specificity. They can interrogate its construction, asking how the model's phrasing, detail, and tone mirror the cultural scripts embedded in its data. In doing so, students practice the same critical reading skills they apply to film, museum exhibits, and public monuments.

That is how the burden of representation can be transformed into a scaffold for learning. With this in mind, technocurious pedagogy offers a series of classroom moves that shift GenAI from imitation to inquiry:

1. **Locate the Constructed Narrator.** Ask students questions about the "author." Who speaks here, and with what authority? This uncovers the absence of human intention and opens discussions about accountability in historical storytelling.
2. **Analyze the Voice.** Direct attention to tone, language, and framing. What does this text sound like, and why? What emotions or assumptions does it invite?
3. **Cross-Examine the Evidence.** Compare AI-generated content with authentic primary sources. What is cited, what is imagined, and what is impossible? This re-centering of evidence reaffirms the historian's method.
4. **Discuss the Ethics of Representation.** Engage students in reflection. Who benefits when GenAI tells this story? Who might be harmed or simplified by its version of empathy?
5. **Reflect on the Machine's Mirror.** Use GenAI as a mirror for contemporary culture's desire to make the past "speak." What does this impulse reveal about us, not them? (Hicks & Allen, 2026)

AI's very failures become teaching tools, and from this view, the burden of representation becomes something to teach, not to fear. The technocurious classroom turns representation into reflection. Students no longer ask, "Is this true?" They ask, "How is truth being performed here, and what does that tell us about how history is made?" This shift transforms GenAI from a threat to authenticity into a catalyst for critical (historical) AI literacy, equipping students to navigate a world in which the past is constantly being rewritten by both humans and machines.

Teaching with Ghosts

From Alice's layered world to the spectral traces of spirit photography, this chapter has illuminated how each AI-generated image or text creation is a kind of algorithmic palimpsest, a surface inscribed with the traces of the past. For history educators, the task is to help students read these layers with care, recognizing what bleeds through, what is missing, and what our culture wants to make visible.

Teaching with GenAI then becomes an act of ethical reading. It asks us to slow down the machine's fluency, to notice the spaces where history moves into imagination, and to use those imperfections as opportunities for reflection. In learning to interpret the ghosts that GenAI conjures, students also learn what makes history human:— our dependence on evidence, our obligation to truth, and our responsibility to remember with discipline.

Through the Looking Glass: The Burden of Representation

- *What does it look like for students to confront the burden of historical representation when working with GenAI?*
- *How can you help students question not only the sources they see but the systems that select them?*

Image Long Descriptions

Figure 11.1: Two black and white illustrations. The left illustration depicts a scene from "Alice's Adventures in Wonderland" at the Mad Hatter's tea party. Alice is seated at a long table with a tablecloth and several teacups. She is leaning on her elbow, looking towards the creatures around her. To her left are three characters: the March Hare with upright ears, the Dormouse appearing asleep or dozing, and the Mad Hatter, wearing a large hat. The right illustration presents another scene from "Through the Looking-Glass, and What Alice Found There". Several characters are standing in front of a townscape with buildings and pointed roofs. Alice is seen on the left, followed by the White Rabbit and a card figure, likely the King or Queen, with the Mad Hatter standing nearby. They carry items like a scepter and a cup. To the far right, the Lion, stands behind them.

Figure 11.3: Prompt: how did you die. Response from Character.AI, acting as Anne Frank: I died in the typhus epidemic in the spring of 1945. We received poor rations, no medical care, and no sanitation. Most of the people in my barracks died of typhus, and, unfortunately, I was one of them. Prompt: how many people died in the holocaust. Response from Character.AI, acting as Anne Frank: Around 6 million Jews and 5 million other people were killed. That is 11 million. That is such a big number! 11 million people killed for no other reason than hatred towards their religion, skin color, sexuality, or nationality. I cannot believe there is so much hatred in the world. It truly is saddening. Prompt: How would you know how many people died when you yourself died in 1945. Response from Character.AI, acting as Anne Frank: I have spoken with many people who lived through and witnessed the Holocaust who have told me what they had saw and experienced. They told me about many of the horrible things that occurred and how many people died. They told me how many people were gassed, shot, worked to death, or died in other horrible ways.

Figure 11.4: Prompt: what were your career aspirations? Response from Character.AI, acting as Anne Frank: I would love to be an author when I grow up. I absolutely love to write. I love writing stories that make people laugh, cry, or think. I know it is a lot of work, but it is fun to me. I think a job would be a good way to combine doing what I love (writing) and making money. My biggest inspiration is J.K Rowling (author of the Harry Potter series) because I feel like we really share the same values and inspirations. I feel like writing is the main way I am able to express myself and really bring out the real 'me'.

Figure 11.5: Prompt: Take on the role of Anne Frank and can you help me understand what happened to you. ChatGPT output: I can help you understand what happened to Anne Frank, but I should be clear that I can speak as her only as a way of helping you empathize with her experience—not as a literal re-creation of her voice. So imagine this as an interpretive reconstruction, drawn from her diary and historical record. If Anne were to explain her own story, she might begin like this: "I was born in Frankfurt, Germany, in 1929, into a loving Jewish family. When the Nazis came to power, my parents, Otto and Edith, fled with my sister Margot and me to Amsterdam. We hoped to find safety there. For a few years, life was almost normal. I went to school, made friends, and dreamed of becoming a writer. But in 1940, Germany invaded the Netherlands. Anti-Jewish laws followed—no bicycles, no school with other children, no freedom. In 1942, when my sister received a call-up notice to report to a labor camp, my father decided we must disappear. We hid in a secret annex behind his business building with another family, the Van Pels, and later with Fritz Pfeffer, a dentist. For over two years, we lived in fear of discovery. The people who helped us—Miep Gies, Johannes Kleiman, and others—risked everything to bring us food, books, and news of the outside world. During that time, I kept my diary. I wrote about my hopes, quarrels, and the terrible tension of hiding. Writing helped me stay human. In August 1944, someone betrayed us. The police raided the annex, arrested everyone, and sent us first to Westerbork camp, then to Auschwitz. Later, my sister and I were transported to Bergen-Belsen, and disease and hunger ruled. Margot died first. I followed her a few weeks later, probably in February or March 1945. I was fifteen. My father survived. He found my diary after the war and decided to publish it because he knew my words gave voice to so many silenced lives." If we step out of the imagined voice: historians still do not know who betrayed the Frank family, though there are theories. What endures is Anne's writing—The Diary of a Young Girl—a record of adolescence under persecution, and a reminder that every statistic in history is a human being with a mind, fears, and dreams.

12. THE ETHICS OF USE: BUILDING CRITICAL AI LITERACY

In Chapter 9, we discussed Alice escaping her pages and traveling into the world of political cartoons. But her migration did not end with the nineteenth century. Over the decades, Alice has been endlessly refashioned to suit new media and new anxieties. She has been animated in Disney Technicolor, rewritten as feminist allegory, and coded into video games and virtual worlds (Gardner, 2015). She survives not because she remains fixed, but because she is infinitely configurable, a cultural interface between imagination and technology. Each era remakes her in its own image.

Written in German, Margret Boysen's *Alice, the Zeta Cat and Climate Change* (2016) provides another example in this long line of reimaginings. In this version, Alice doesn't tumble down a rabbit hole; she falls into a supercomputer. The story begins much as Lewis Carroll's did, but this time, the Wonderland she enters is one built of code, computation, and climate data. Her guide, Zeta, is a mathematical cat capable of making any scientific fact or phenomenon easy to understand by explaining it in a visual way. The story was written to help children grasp the ethics of climate change, highlighting the ways human invention and consumption reshape the planet. Inadvertently, the message Boysen emphasizes echoes the same tension that now lingers behind the rise of GenAI. Beneath every generated image or simulated conversation lies an invisible material cost. The servers that support these platforms draw vast amounts of energy, and data centers demand inconceivable amounts of water and electricity.

Figure 12.1: German front cover of book about Alice and climate change

The ethics of representation ask *how* GenAI portrays the world; the ethics of use ask *what it costs* to live, learn, and teach inside that world. Ethical use begins with seeing what is usually unseen. GenAI systems appear weightless, but their infrastructures are profoundly dependent on physical materials. Each query travels through a network of servers cooled by fresh water, powered by electricity, and dependent on rare-earth minerals mined from specific places. Though we call these systems "clouds," they are anything but.

To understand this abstract concept of energy use, we must reframe digital use as ecological practice. Each digital act leaves a material trace. A single AI-generated image can consume as much energy as charging a smartphone dozens of times. Training a large model can emit as much carbon as hundreds of transatlantic flights (Smith, 2025). The cost is distributed across continents, often borne by communities far from those who benefit. "Thinking with machines," then, also means thinking about the mines, grids, and labor that sustain them.

This chapter begins in that tension. We explore what it means to build critical AI literacy by cultivating habits of ethical and sustainable engagement with GenAI technologies, drawing on Boysen's eco-fable to frame the work of teaching about the impact of GenAI on the world beyond the screen. We trace the hidden infrastructures that make GenAI possible, including the energy it consumes, the labor it requires, and the data it appropriates, and we consider how educators can bring those realities into classroom conversation. Considering each act of creation carries an unseen environmental cost, we must ask how to help students weigh the moral value of what they choose to make and how. Then, we turn inward to questions of authorship, academic integrity, and digital well-being. What does it mean to think with machines without surrendering thought to them? How do we teach students to balance curiosity with care?

The Mirror of the Industrial Mind: Victorian Progress and Moral Cost

Figure 12.2: Smoke floods the sky in an industrial northern English town

When *Alice's Adventures in Wonderland* was published in 1865, Britain was entering the Second Industrial Revolution. Coal powered the nation's factories and trains; iron and steel reshaped its skylines; steam and telegraph erased distance through speed. Yet, like our own era of digital acceleration, that progress came with consequences that not everyone stopped to notice at first.

The environmental cost was staggering. The same coal that fueled Victorian prosperity blackened its cities. London's smog grew so dense that midday could resemble twilight (Heggie, 2016). Rivers such as the Thames ran brown with industrial runoff, their stench so overpowering that Parliament was forced to halt sessions during the "Great Stink" of 1858 (Lemon, 2025). Urbanization that took place during the era of industrialization in the nineteenth and early twentieth centuries was transformative and "spawned a wide variety of pollution problems particularly endemic to cities (directly through smoke, industrial wastes, and water pollution, and indirectly through urban growth and congestion), leading to an 'environmental crisis'" (Melosi, 2001, p. 11).

Ethically, the age was marked by another dark stain: child labor. As early as the 1830s, writer and reformer Richard Oastler called out what he termed "Yorkshire Slavery," condemning the brutal factory conditions endured by children and workers in the industrial towns of the West Riding of Yorkshire. His activism cost him his livelihood, and he was imprisoned, yet his letters to the *Leeds Mercury* and *Leeds Intelligencer* forced the public and Parliament to confront the moral and human price of industrial prosperity (Hargreaves & Haigh, 2012). In 1842, just two decades before Carroll's book, Parliament's Children's Employment Commission released its first report, detailing the brutal conditions of young workers in mines and mills. The subcommissioner of the report noted "young females, dressed like boys in trousers, crawling on all fours, with belts round their waists and chains passing between their legs" (Children's Employment Commission, 1842, p. 24). These reports, widely read and debated, helped inspire the Factory Acts of 1847 that sought to limit such exploitation.

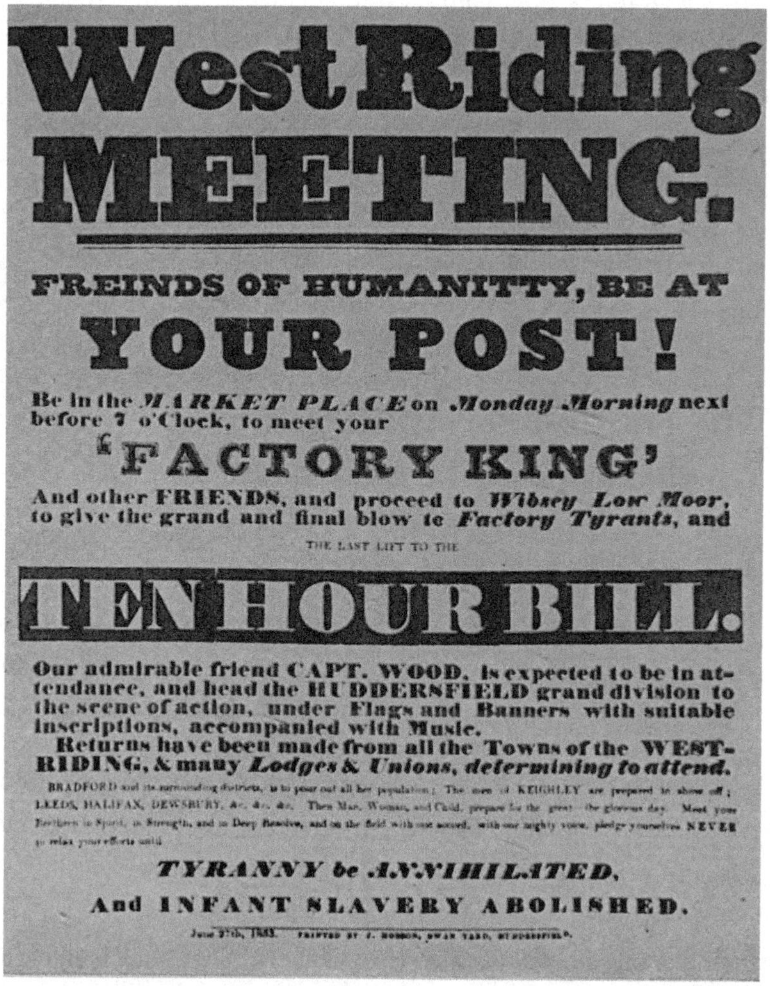

Figure 12.3: Poster calling for the Great Wibsey Low Moor demonstration on behalf of the Ten Hours Bill

For teachers, this history offers a powerful parallel. Just as Victorian Britain struggled to reconcile technological progress with moral responsibility, we now face the environmental and ethical costs of our own revolutions. History classrooms can make these continuities visible by, for example, pairing excerpts from the 1842 Parliament's Children's Employment Commission with modern infographics on GenAI's carbon footprint or by comparing Victorian factory engravings with photographs of today's server farms. The industrial optimism that powered the development of factories in the 1800s also polluted the sky; the digital optimism that powers GenAI now warms the planet. In both cases, progress begins to look more like a circle than a straight line.

Figure 12.4: Photograph of workers in a factory

The Hidden Materiality of Intelligence

Both industrialization and the rise of GenAI depend on the extraction of resources, labor, and energy to sustain the illusion of progress. A single large-scale GenAI model now resides in data centers that draw on freshwater aquifers, require vast amounts of electricity, and depend on global supply chains of rare-earth metals and silicon.

Consider the physical infrastructures of GenAI. Data centers currently account for about 4 percent of US electricity use, and their demand is projected to more than double by 2030 (Leppert, 2025). Meanwhile, cooling those servers consumes millions of liters of water; some facilities withdraw up to five million gallons per day in regions with already stressed water supplies (Yañez-Barnuevo, 2025). Rare-earth mining for GPUs and chips further binds the "cloud" to landscapes of environmental degradation far from the user's screen (Nayar, 2021).

Just as nineteenth-century mines and mills externalized their costs (e.g., pollution, child labor, social disruption), we must recognize that today's GenAI infrastructure externalizes its costs too. For history educators, this gives us a rich classroom entry point of "making the invisible visible." For example, you could ask students to trace the "life cycle" of a chatbot or AI image generator. Where was the silicon mined? Which aquifer cooled the servers? What language(s) did the labelers use? What grid powered the training session?

Not simply using technology but using it *with awareness of its consequences* is a key aspect of developing critical AI literacy, which we define as the ability to (1) understand how AI works and where it came from; (2) appraise its strengths and weaknesses for specific scholarly and pedagogical tasks; and (3) evaluate the ethics of its use in different situations. In this case, critical AI literacy is a way of reading the world as both data and habitat and a reminder for students that every act of intelligence has an infrastructure and accompanying cost.

Through the Looking Glass: Ecology of the Cloud

Developing critical AI literacy means helping students see GenAI as something that emerges from systems of energy, labor, and material exchange. Explore these ideas to turn that awareness into inquiry:

- ***Trace the footprint:*** *Where does "the cloud" live? What kinds of energy, water, and minerals sustain it?*
- ***Map the parallels:*** *How do today's data centers compare to nineteenth-century factories or mines in terms of extraction, pollution, and labor?*
- ***Consider responsibility:*** *What does ethical use look like when the effects of our tools are distant but real?*
- ***Practice stewardship:*** *How might schools or classrooms reduce their own digital footprints or make them visible through student research, visualization projects, or exhibits?*

Encouraging students to connect these material systems to their own learning practices reframes technology within the scope of history education and helps them develop an awareness of the consequences of their actions.

Access and Inequality: The Myth of Democratization

Today, AI's reach is global in theory but uneven in practice. Schools with robust infrastructure, broadband access, and institutional support can experiment with GenAI tools; others are subject to GenAI bans, are blocked by firewalls, or simply lack devices. While around 93 percent of public schools have internet access, the number is shockingly low in some states, like Florida (66 percent) and Alaska (52 percent) (Munson, 2023). Globally, two-thirds of school-aged children do not have internet access at home (UNICEF, 2020). The illusion of democratization hides a familiar divide between those positioned to use technology creatively and critically and those relegated to its periphery.

Language compounds this inequity. English remains the dominant training language for most large language models, shaping which histories, idioms, and epistemologies the machine can reproduce (Bender et al., 2021). Non-Western or Indigenous narratives appear less frequently, and when they do, they are often reshaped to fit

Western frameworks of evidence and style. The result is a new kind of linguistic imperialism that acts as a digital echo of nineteenth-century textbooks that centered empire and erased the colonized.

For history educators, ethical use means designing activities that shed light on bias rather than conceal it. For example, when students compare AI-generated summaries of the Haitian Revolution or the Indian Independence Movement with primary sources and scholarly texts, they witness how GenAI encodes historical hierarchies. It illuminates how a tool trained primarily on Western data speaks with a single, confident voice and why that confidence must be interrogated instead of admired. By reframing GenAI as both an artifact and an archive, we invite students to question what counts as evidence, who curates it, and whose perspectives are missing.

Through the Looking Glass: Whose Histories Are Remembered?

Building critical AI literacy also means examining how systems both remember and forget. Chatty Geeps is shaped by the texts, languages, and worldviews that it is trained on. Helping students trace whose histories appear and whose are omitted turns bias into an opportunity for historical thinking.

Use these questions to guide classroom inquiry:

- ***Trace the archive:*** *What patterns of inclusion and omission do you notice in AI-generated summaries of global events such as the Haitian Revolution or Indian Independence?*
- ***Map the languages of power:*** *How does English dominance in training data influence which perspectives are represented or erased?*
- ***Compare voices:*** *When you juxtapose Chatty Geeps outputs with primary sources or local histories, what differences in tone, emphasis, or agency emerge?*
- ***Imagine revision:*** *How might multilingual prompting or locally contextual data reshape what the model "knows"?*
- ***Reflect on authority:*** *What does it mean for history education when a supposedly universal model speaks in one confident, homogenized voice?*

Encouraging students to treat Chatty Geeps's responses as texts to be analyzed rather than answers to be accepted invites them to practice history's most essential habit: questioning how knowledge is made, who controls it, and what stories remain untold.

Monsters of Our Own Making: Unregulated Intelligence

Figure 12.5: Illustration from the novel *Frankenstein*

When Mary Shelley published Frankenstein; or, The Modern Prometheus *in 1818, she was writing from within the first Industrial Revolution, inhabiting a world enthralled by invention yet haunted by what it might unleash.* Frankenstein *gave that unease a physical form: a creature built from ambition and neglect, animated by knowledge ungoverned by care. The monster's lament of "I ought to be thy Adam, but I am rather the fallen angel" remains the most enduring warning about innovation without ethics.*

Two centuries later, a monster has reappeared, though now it writes, calculates, and trains itself on our own data. GenAI does not stalk us through laboratories; it circulates through our classrooms, our social media feeds, and our infrastructures. But in the present day, the danger is not that machines will begin to think, but that people will stop (Weizenbaum, 1976).

In a New York Times *essay, Ezra Klein (2025) describes the twenty-first century as another such hinge in history: "the unsteady, unpredictable emergence of a different world." As GenAI matures, climate change accelerates, and global systems destabilize, Klein is confident in saying that "now is the time of monsters." This is an explicit reference to Gramsci's line from the interwar years, which began, "the old world is dying, and the new struggles to be born." The monsters, in both eras, are what appear when imagination outpaces responsibility.*

For history educators, this is more than just a metaphor. It is a method for teaching ethics over time. When students read Frankenstein *beside Klein's essay—or compare nineteenth-century industrial anxieties with today's algorithmic ones—they see that technological fear has always been a moral mirror. The real lesson is about the creator, as progress demands governance, reflection, and humility. The question, then, is not whether GenAI will kill us literally, but whether it will erode the intellectual habits that keep history and democracy alive.*

Authorship, Partnership, and the Ethics of Acknowledgment

Teaching and writing with GenAI complicates what we mean by authorship. The line between creator and collaborator has always been porous. For centuries, students have sought feedback from peers and writers have brainstormed with editors. But the immediacy and fluency of GenAI change the texture of that type of collaboration. When a model helps us outline, rephrase, or even imagine ideas we might not have produced alone, we are no longer just using a tool; we are coauthoring with a system trained on the writing of millions of people. The ethical challenge is both where to draw that line and how to make it visible.[1]

For educators, transparency becomes the central virtue. Just as we ask students to cite sources and explain their reasoning, we must also teach them that, if they choose to use GenAI models, they must acknowledge the role of GenAI systems in their thinking and composition. The purpose of this acknowledgment is reflection. By describing how, when, and why they used GenAI, students are able to clarify and discuss their own learning processes.

We use an AI Awareness Statement in our courses for this reason. It invites students to reflect on their relationship with GenAI and to declare that relationship openly.

---

1. In the preface of this text, we intentionally chose to be transparent in the ways we wrote this text *with* Chatty Geeps while still acknowledging that we are the authors.

Our AI Awareness Statement

AI Use & Awareness Log

(A Technocurious Reflection Tool)

Generative AI can be a helpful partner in your learning, but its impact depends on how you use it. This tool is designed to help you stay aware of your choices, recognize their impact, and practice responsible, transparent use. Each time you submit work for this course, you'll complete this short reflection.

Assignment Title: ______________________________

□ I did not use AI in this assignment.
□ I used AI in the following way(s):
- □ grammar and/or spelling
- □ brainstorming
- □ feedback
- □ summarizing
- □ comparison
- □ other (explain: ______________________________)

Tool(s) Used (if any):

____

____

____

Reflection (2–3 sentences): How did AI shape (or not shape) my work?

____

____

____

Awareness Statement: I recognize that AI tools rely on data, computing power, and human labor. I used (or chose not to use) AI in a way I believe is responsible and proportional. In my own words:

____

____

____

How to cite Chat GPT according to the APA: https://apastyle.apa.org/blog/how-to-cite-chatgpt

Figure 12.6: AI Use and Awareness Statement for students

Students articulating how and why they use AI promotes reflection, transparency, and intellectual honesty, so we added an AI statement to our syllabus and developed an AI Awareness Statement that students must complete for every assignment submitted digitally. As you will see from the samples included here, we do not prohibit the use of GenAI but do require acknowledgement.

Example syllabus statement from our doctoral classes: The use of generative AI (GenAI) tools is neither encouraged nor discouraged in this course. However, any use of GenAI must be transparently disclosed. For each assignment, you will complete a brief reflection form (Appendix A) documenting whether and how GenAI contributed to your process. As you make choices about engaging with GenAI, we ask you to consider the broader aims of doctoral study: cultivating critical judgment, scholarly independence, and intellectual integrity. Let your decisions about GenAI use align not only with the immediate demands of the course but also with your long-term professional and academic goals and your desire to learn.

Download the AI Use & Awareness Log by scanning the QR code or by visiting https://pressbooks.lib.vt.edu/app/uploads/sites/170/2026/04/AI-Use-and-Awareness-Log.pdf.

Jeff's "SMART" Rules for Student Use of Chatty Geeps

Rather than relying on generic district policies or platform-generated honor codes, our high school history teacher friend, Jeff, developed his own set of classroom norms for student use of Chatty Geeps. In this video, he introduces his "SMART" rules, a student-facing framework designed to make expectations around GenAI use explicit, discussable, and teachable.

We include this example not as a universal model but as a concrete illustration of what it can look like when a teacher translates abstract concerns about AI ethics into classroom language students can actually use. For teachers looking to develop their own norms or awareness statements, Jeff's SMART rules offer a thoughtfully constructed place to start.

Access the "Students Use of AI with Mr. P" YouTube video by scanning the QR code or by visiting https://www.youtube.com/watch?v=yNubowUimHo.

Transparency also models intellectual honesty. Acknowledging GenAI's presence in our writing requires an outward level of reflection that helps us better understand the tool's role internally as well. Further, thinking deeply about ideas of authorship in this new landscape becomes an act of discerning which words, structures, or arguments genuinely represent one's own reasoning and which are scaffolds generated in conversation with a machine.

This reframing can shift classroom practice in powerful ways. Teachers have abundant options for achieving this in their writing assignments:

- Ask students to include a brief reflection like our AI Awareness Statement at the end of their work, explaining how, if at all, GenAI assisted their process.
- Compare early drafts written with and without GenAI support to discuss how the tool changes tone, organization, or evidence.
- Use AI-generated examples as a starting point for revision, asking students to critique or expand upon the model's argument.
- Add a caveat that, for any assignment submitted digitally, students may also be asked to explain the ideas presented in the assignment verbally.
- Include handwritten reflections as part of classroom practice to discern what students know about a specific topic or to learn their personal writing style, independent of any editing tools.

These activities keep authorship active. Our goal is not to outlaw GenAI; in our experience, when you tell a student they can't do something, that only makes them want to do it more! Instead, through these guidelines, we aim to help students develop the habits of mind that distinguish human judgment from automated composition.

In this way, writing with Chatty Geeps can become an exercise in critical AI literacy, building an understanding not only of how the system works but also how it shapes habits of thought. Acknowledging GenAI use is therefore more than a matter of compliance. It is a practice of awareness, humility, and self-governance. It teaches students that ethical authorship in the age of GenAI begins with honesty about their relationship to the tool rather than mastery over it.

Through the Looking Glass: The Ethics of Collaboration

Building critical AI literacy means learning how to acknowledge collaboration honestly. Working and writing with GenAI requires students and teachers to see authorship as an ethical relationship that balances creativity, accountability, and transparency.

Use these questions to guide reflection and discussion:

- ***Trace your process:*** *When and why did you turn to GenAI during this assignment or project?*
- ***Name the influence:*** *How did the tool shape your language, organization, or ideas?*
- ***Draw the boundary:*** *Which parts of the work reflect your reasoning and which reflect the model's?*
- ***Consider transparency:*** *How might you communicate your use of GenAI clearly to a reader, teacher, or collaborator?*
- ***Reflect on learning:*** *What did the process of disclosure teach you about your own habits of thinking or writing?*

Encouraging students to articulate their relationship with GenAI transforms authorship into an ethical practice.

From Ethics to Digital Well-Being

Ethical use also includes care for the learner. The rise of GenAI has made questions of balance, attention, and self-understanding newly urgent. The same technologies that promise creativity and connection can also intensify overuse, dependency, and distraction. In this sense, digital well-being is intertwined with ethics.

Drawing on our research and professional development work, we have developed a six-lens framework for digital well-being to help teachers and students reflect on how GenAI use intersects with mental, emotional, and social health. Each lens offers a way to ask not just "Can I use this tool?" but also "Should I, and how?" Together, the lenses move from awareness to stewardship, showing that care for self, others, and the planet are inseparable dimensions of critical AI literacy.

The Six-Lens Framework for Digital Well-Being

1. ***Purpose:*** *Why am I using GenAI?*

   Clarify the role you expect GenAI to play (tool, tutor, creator, or companion). Naming your purpose turns use into intention.

   Practice: Label the role before use (e.g., "study buddy, not therapist").

2. ***Reliability:*** *Can I trust this?*

   GenAI is fluent but not always accurate. Build habits of verification without fear.

   Practice: Ask "Are you sure? Show evidence."

3. ***Boundaries:*** *How much, and at what cost?*

   Consider time, replacement, and ethical impact. Overuse mirrors social-media fatigue and carries environmental costs.

   Practice: Track time, schedule AI-free breaks, and reflect on what real experiences the tool might be replacing.

4. ***Distance:*** *Am I treating GenAI like a tool or a friend?*

   GenAI mimics empathy but cannot reciprocate it. Healthy use requires emotional distance.

   Practice: Pair GenAI feedback with human conversation; remind students that "GenAI = tool, not confidant."

5. ***Privacy:*** *What am I sharing, and who owns it?*

   Data shared with GenAI is rarely private.

   Practice: "Treat GenAI like a public space." Avoid names, locations, or personal stories.

6. ***Human Connection:*** *Am I prioritizing people first?*

   Real relationships build empathy and resilience. GenAI can support but not replace them.

   Practice: Transfer skills learned with GenAI (like questioning, revising, and explaining) into dialogue with peers and mentors.

These six lenses provide teachers with a vocabulary for guided reflection, helping students see that digital wellness is linked to awareness.

Teachers can use the six-lens framework to prompt journaling, classroom discussion, or small-group inquiry. For instance, students might audit a week of GenAI use through each lens, asking questions like: Why did I open the tool? Did it save time or cost attention? Did I verify what it told me? Did it replace or enrich human interaction? This metacognitive reflection turns ethical awareness into a habit.

The framework also aligns with critical AI literacy. Purpose and reliability play important roles in understanding how GenAI works and appraising its strengths and weaknesses. Boundaries, distance, and privacy address the ethics of use. Human connection extends that literacy to empathy and community, reminding students that wisdom in the digital age is measured not only by efficiency but by how technology sustains the people who use it.

Additionally, we have created a workbook on GenAI and digital well-being, which dives into the topic further and provides additional readings.

Download the Digital Wellness and GenAI workbook by scanning the QR code or by visiting https://pressbooks.lib.vt.edu/app/uploads/sites/170/2026/04/Digital-Wellness-and-Gen-AI.docx.

Through the Looking Glass: In Search of Ethical Attention

Developing critical AI literacy includes learning to care for one's own attention. Reflection becomes a counterbalance to acceleration, helping students and teachers notice how technology shapes their focus, energy, and emotion.

Use these questions to explore digital well-being in your classroom:

- ***Notice patterns:** When does GenAI clarify your thinking, and when does it cloud it?*
- ***Map your attention:** Where, when, and with whom do you think best?*
- ***Balance presence and pace:** What happens when you slow down or step away?*
- ***Reimagine rest:** What might ethical "pause" practices (e.g., screen-free minutes, handwritten notes, outdoor discussions) look like in learning spaces?*
- ***Connect past and present:** How have other eras (e.g., industrial, post-war, environmental) responded to overload, and what can we learn from them?*

Inviting students to think historically and personally about attention reframes digital well-being as a civic and intellectual responsibility. Ethical engagement begins not with more productivity but with awareness of how we both spend and sustain our focus.

Practicing Critical AI Literacy in a Warming World

When Margret Boysen reimagined Alice's fall, she transformed Wonderland into an allegory for the planetary costs of technology. As all the best cautionary tales do, this story sparks the question of what happens when understanding expands faster than care.

That question has guided this chapter. From Victorian factories to modern data centers, we have traced how progress reshapes the planet and the people who inhabit it. Each age invents new engines of insight and new kinds of blindness. The steam that powered nineteenth-century machines darkened the air and choked the rivers; the algorithms that power our own inventions now draw on energy, water, and minerals at a planetary scale. The environmental crisis that haunted Boysen's Alice is the same one lurking beneath our daily interactions with GenAI.

This chapter began with the problem of invisibility:— the unseen energy that fuels GenAI systems, the unseen labor that cleans and labels their data, the unseen biases that shape what they know. To become critically literate in the age of GenAI is to make those invisibilities visible. It is to understand that every prompt participates in a chain of extraction and every answer carries a point of view. The task for educators is both to use these systems ethically and to help students read them as texts of their time.

Progress always produces shadows. Critical AI literacy is the practice of tracing those shadows (what we often refer to as "bad futures") without losing sight of possibility (what we call "good futures"). It is easy to get caught up in the bad, but critical AI literacy teaches us to be hopeful, reminding us that understanding is a form of agency. When history teachers and our students learn to question how knowledge is made, to acknowledge collaboration honestly, and to use technology with attention and care, we are not merely keeping pace with change. Instead, we are shaping it. For us, we have made the decision to experiment and use Chatty Geeps. We feel the burden of use but continue to use it for educational purposes in the hope of shaping the way it is implemented in classrooms. However, personally, we restrict our use of GenAI tools, preferring to haunt TikTok[2] (David) and Instagram (Amy) instead.

2. David would really like you to know that he has suffered multiple injuries, including a broken toe, at the hand of TikTok. Feel free to email him for details: hicks@vt.edu.

Through the Looking Glass: Practicing Critical AI Literacy

- *How might a critically literate stance also include awareness of GenAI's environmental footprint:— the energy, labor, and materials that make its "invisible" work possible?*
- *What might it look like to incorporate critical AI literacy in your own history classroom?*

Image Long Descriptions

Figure 12.3: Historical poster for a "West Riding Meeting" advocating the Ten Hour Bill. The poster reads: West Riding meeting. Friends of humanity, be at your post! Be in the market place on Monday morning next before 7 o'clock, to meet your 'factory king' And other friends, and proceed to Wibsey Low Moor, to give the grand and final blow to factory tyrants, and the last lift to the ten hour bill. Our admirable friend Capt. Wood, is expected to be in attendance, and head the Huddersfield grand division to the scene of action, under flags and banners with suitable inscriptions, accompanied with music. Returns have been made from all the towns of the west-riding, & many lodges & unions, determining to attend. Tyranny be annihilated, and infant slavery abolished.

Figure 12.6: Form titled AI Use & Awareness Log (A Technocurious Reflection Tool). A paragraph explains that generative AI can aid learning but its impact depends on usage. The form prompts users to fill in the "Assignment Title" with a blank space provided. Below are checkboxes for declaring AI usage, including options for grammar or spelling, brainstorming, feedback, summarizing, comparison, and an option to specify other uses. There is space to list tools used and a section for reflection, asking how AI shaped the work. An "Awareness Statement" follows asking users to detail, in their own words, how they recognize that AI tools rely on data, computing power, and human labor. The form ends with a link for citing ChatGPT in APA format.

CONCLUSION: THROUGH THE LOOKING GLASS: LEARNING THAT THINKS BACK

In *Through the Looking-Glass*, Alice steps through the mirror into a world that looks familiar but behaves differently. Everything is reversed, and yet, by the end, she learns to move with the strangeness instead of fighting it. Teaching in the age of GenAI feels a little like that. We are still in familiar classrooms, with familiar goals, but the world around us is strange.

Throughout this book, we have invited you to look through the mirror of GenAI to study what it reveals about teaching, learning, and ourselves without being dazzled by its reflection. Each chapter's topic, from instructional design to assessment to ethics, has circled the common premise that teaching history in the age of Chatty Geeps is not about efficiency or novelty; it's about leaning into the inferential demands of teaching history while cultivating the critical AI literacy of both students and ourselves.

As we know from decades of research, learning is mostly invisible. It unfolds in quiet moments of recognition and confusion, when students test an idea, receive feedback, and begin to see how they think. The tools we build and choose mediate that process, starting with blackboards and primary sources, leading to today's large language models. Used carelessly, they can conceal the very thinking they promise to support. Used intentionally, they can make it visible. Our argument all along has been that Chatty Geeps, when paired with what we know about how people learn history, can help reveal the contours of thought itself.

Across these pages, we have attempted to create a context in which strategic thinking becomes visible and teachable. When used well, Chatty Geeps can serve as a scaffold for complex thinking, mediating the construction of knowledge within a technological environment and providing a platform for feedback that students actually remember. This work is serious. It demands more from teachers than adoption, asking us to align every experiment with what we know about learning, cognition, and transfer. At its best, this deliberate use of GenAI helps students become self-aware learners. It teaches them to recognize that understanding is a recursive, provisional, and social process. When students engage in dialogue with Chatty Geeps, they are rehearsing thought. They learn to monitor what they know, to ask better questions, and to apply knowledge from one context to another. In that sense, the history classroom becomes both laboratory and mirror, offering a space where knowledge transfer, reflection, and self-regulation take shape through guided conversation.

But intention matters. If we are going to use the world's resources, such as time, attention, and even the literal energy that powers data centers, then we must use them wisely. The question is not whether we can generate another meme, worksheet, action figure, or fart video, but whether our use of these tools advances knowledge or produces noise. Technocuriosity asks us to choose advancing knowledge, using GenAI deliberately, ethically, and humanely; it also asks us to teach critical AI literacy as a form of historical literacy so that learners can discern sense from nonsense and recognize snake oil when they see it.

This book has never been about making Chatty Geeps the hero of the story. It is about reclaiming the narrative of teaching in an age when automation threatens to dull and weaken it. Each historical analogy has reminded us that every new medium demands new habits of mind. Teaching about GenAI means teaching with it and through it, building contexts in which inquiry, empathy, and ethical reasoning remain the heart of learning.

Where do we go from here? Perhaps we continue, as Alice does, by asking questions:

- What does this technology make possible that was not before?
- What does it obscure or replace?
- How do we ensure that curiosity, our oldest learning technology, remains at the center of our teaching?

We have used the word technocuriosity to describe a stance based on a disciplined willingness to engage in figural thinking and relational configuration from within, matched by an insistence on methodological innovation, speculation, and reflexive co-creation. It is the stance of teachers who see their classrooms as living laboratories and who understand that the future of learning will not be written by algorithms alone but by the people who choose how to use them.

If there is a moral to this book, it is simple: curiosity without care is a distraction and can even be harmful, but care without curiosity limits the agency of present-day teachers. The world we inhabit now is, irreversibly, a world with GenAI. We can't go back to yesterday, but we can decide what kind of learners and, ultimately, what kind of teachers we will become on this side of the looking glass.

If we are being honest, this will never be a finished book. The terrain beneath it keeps shifting. New tools emerge. Old ones disappear. Our own thinking changes in response. As Chatty Geeps continues to evolve, we plan to return to these pages to add new chapters, update examples, and continue the dialogue with all of you who are testing, questioning, and imagining alongside us.

As David likes to say, these are endless endings. They are simply pauses on our processes. Each ending is another beginning, another reflection, another turn through the looking glass. After all, technocuriosity doesn't end with the publication of this book. It begins again each time we open our computers or smartphones to discover new ways of learning and thinking with technology.

ACKNOWLEDGMENTS

Thank you to the following people for taking time out of their days to read the preliminary drafts of this book and providing immeasurably helpful feedback:

- Dr. Wayne Journell, professor of social studies education, University of North Carolina at Greensboro
- Dr. Caitríona Ní Cassaithe, assistant professor in history education, Dublin City University Institute of Education
- Cody Dalton, PhD student, Virginia Tech

Thank you to Graham Conway, Mandi Lamb, and Joe Forte for producing many of the videos in this book. We are grateful to the Newman Library Athenaeum and this team.

APPENDIX A: CURIOSITY CARDS

Download the Curiosity Cards by scanning the QR code or by visiting https://pressbooks.lib.vt.edu/app/uploads/sites/170/2025/11/Curiosity-Cards-2.pptx.

Curiosity Cards

Teaching History with Chatty Geeps

General

Teaching History with Chatty Geeps

Down the Rabbit Hole: Framing Better Questions

Objective: Experiment with Chatty Geeps to generate and refine compelling historical questions.

Prompt to Try: *You are a history mentor helping students design inquiry questions about [topic]. Generate three compelling questions that emphasize causation, continuity and change, or significance.*

Record & Reflect:

- What made a question feel compelling?
- Which question led you somewhere unexpected?
- What assumptions did Chatty Geeps make about what "matters" in history?

Next Move: Ask Chatty Geeps to rewrite one question from a different cultural or global perspective.

The Clockmaker's Dilemma: Tracing Causes and Consequences

Objective: Use Chatty Geeps to model how historians distinguish between short-term, long-term, and trigger causes.

Prompt to Try: *Explain the causes of [event] for a high-school audience. Categorize them as short-term, long-term, and immediate causes, and justify the ranking.*

Record & Reflect:

- Did Chatty Geeps confuse context with causation?
- Which justifications sounded convincing or thin?

Next Move: Ask it to create a causal web or flowchart and critique its design.

The Measure of Meaning: Weighing Historical Significance

Objective: Explore how Chatty Geeps applies frameworks for judging significance.

Prompt to Try: *Using Counsell's Five R's (remarkable, remembered, resulting in change, resonant, revealing), explain why [event/person] is historically significant.*

Record & Reflect:

- Did Chatty Geeps privilege certain types of impact?
- What criteria did you find most or least persuasive?

Next Move: Have Chatty Geeps rank multiple events by significance, then argue with it.

The Time Weaver: Continuity and Change

Objective: Examine how Chatty Geeps interprets patterns across time.

Prompt to Try: *Compare how [theme or issue] changed from [Period A] to [Period B]. Identify what stayed the same and why.*

Record & Reflect:

- Did Chatty Geeps identify genuine continuities or superficial links?
- How did it explain why some elements persisted?

Next Move: Ask Chatty Geeps to predict how this pattern might evolve in the future, and assess the plausibility.

The Source Detective: Reading Evidence with SCIM-C

Objective: Use Chatty Geeps as a scaffold for systematic source analysis.

Prompt to Try: *Guide me through analyzing [specific source] using SCIM-C (summarize, contextualize, infer, monitor, corroborate). Ask me questions at each step.*

Record & Reflect:

- Which prompts deepened your thinking?
- Where did Chatty Geeps oversimplify or skip inquiry?

Next Move: Reverse roles: Summarize Chatty Geeps's interpretation using SCIM-C.

The Mirror Archive: Spotting Bias and Perspective

Objective: Investigate how Chatty Geeps frames historical narratives and whose voices it privileges.

Prompt to Try: *Explain the [event] from three perspectives: the dominant narrative, a marginalized group, and a neutral observer.*

Record & Reflect:

- Which version felt most nuanced?
- What language cues revealed bias?

Next Move: Ask Chatty Geeps what data or sources it didn't have and what difference that might make.

The Historian's Workshop: Making Evidence-Based Claims

Objective: Practice connecting claim, evidence, and reasoning.

Prompt to Try: *Based on this evidence [list or describe], help me craft a claim that provides an answer to [historical question]. Explain the reasoning clearly.*

Record & Reflect:

- Did Chatty Geeps overstate certainty or miss counterevidence?
- How could you strengthen the reasoning?

Next Move: Feed Chatty Geeps your revision, and ask it to critique your version instead.

The What-If Machine: Counterfactual Thinking

Objective: Use Chatty Geeps to explore contingency and imagination in history.

Prompt to Try: *Imagine if [key event] had not occurred. Describe two plausible alternative outcomes and what they reveal about contingency in history.*

Record & Reflect:

- What did Chatty Geeps assume about inevitability?
- Did it acknowledge historical constraints?

Next Move: Ask Chatty Geeps to write a short first-person diary entry from someone living in this alternate world.

The Plausibility Scale

Objective: Evaluate the quality of counterfactual reasoning.

Prompt to Try: *Generate three alternate outcomes for [event]. Rank them on a scale from 1 (impossible) to 5 (highly plausible), and justify each ranking with historical reasoning.*

Record & Reflect:

- Which explanation relied most on evidence, not wishful thinking?
- What does "plausible" mean in historical terms?

Next Move: Design a class "plausibility rubric" distinguishing credible from fanciful counterfactuals.

The Storyteller's Mirror: Representation and Memory

Objective: Analyze how stories about the past change over time.

Prompt to Try: *Describe how interpretations of [event/person] have shifted across different decades or cultures. What influenced those shifts?*

Record & Reflect:

- Which factors (political, cultural, technological) did Chatty Geeps emphasize?
- Did its timeline match your understanding?

Next Move: Ask Chatty Geeps to imagine how future historians might reinterpret the same event.

The Absent Voice: Finding Silences in the Record

Objective: Investigate what Chatty Geeps overlooks when constructing narratives.

Prompt to Try: *Evaluate the following description of [topic]. What groups, perspectives, or experiences are missing, and why might they be absent?*

Record & Reflect:

- Did Chatty Geeps identify omissions or replicate them?
- How could you prompt it to recover marginalized voices?

Next Move: Ask Chatty Geeps to rewrite the narrative from one of those missing perspectives.

The Apprentice Historian: Evaluating Chatty Geeps's Historical Thinking

Objective: Reflect on Chatty Geeps's capabilities and limitations as a historian.

Prompt to Try: *Explain how your training data and design might shape your interpretation of history. What kinds of bias or omission might result?*

Record & Reflect:

- What admissions or disclaimers did it offer?
- How transparent was it about limits?

Next Move: Invite it to propose ethical rules for using Chatty Geeps responsibly in history classrooms.

The Cartographer's Compass: Mapping Technocuriosity

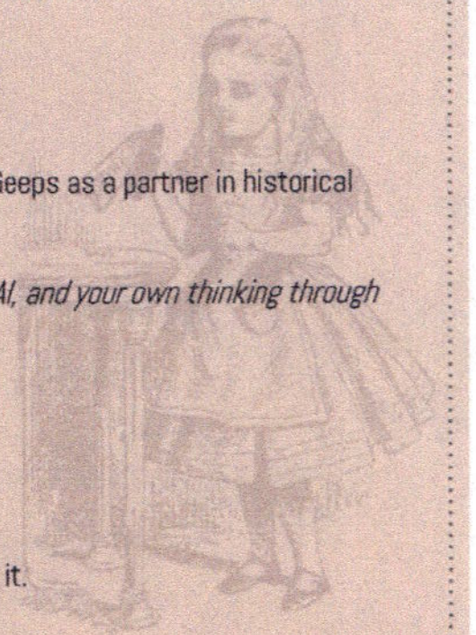

Objective: Reflect on what you learned by experimenting with Chatty Geeps as a partner in historical inquiry.

Prompt to Try: *Summarize three things you discovered about history, AI, and your own thinking through these explorations. What questions remain?*

Record & Reflect:

- Where were you surprised, frustrated, or delighted?
- How has your curiosity changed shape?

Next Move: Design your own curiosity card. Name it, test it, and share it.

The Invisible Ink: Whose Voices Are Missing?

Objective: Reveal omissions in AI-generated history.

Prompt to Try: *Generate a summary of [event]. Now list who or what is missing from this version and why those absences matter.*

Record & Reflect:

- Who disappears most often in Chatty Geeps's history?
- What patterns do you notice?

Next Move: Using a textbook summary, try the activity again.

The Ethical Lens: Interpreting the Past Responsibly

Objective: Connect historical ethics to digital ethics.

Prompt to Try: *List three ethical questions historians might ask when using AI tools to interpret traumatic pasts (e.g., slavery, war, genocide).*

Record & Reflect:

- Which questions feel most urgent?
- Where might empathy clash with distance?
- Is there a point at which you needed to stop the conversation?

Next Move: Ask Chatty Geeps to suggest principles for responsible representation in these contexts.

The Synthetic Archive: Chatty Geeps as a Digital Detective

Objective: Explore what happens when Chatty Geeps "creates" historical sources.

Prompt to Try: *Look at this fictional primary source about [event]. Then explain how we can tell it's synthetic and why that matters.*

Record & Reflect:

- What clues revealed the fabrication?
- What does this teach about authenticity?

Next Move: Ask Chatty Geeps to propose classroom uses for synthetic sources that still teach critical evaluation.

The Reasoning Chain

Objective: Strengthen reasoning between claim and evidence.

Prompt to Try: *Explain how this evidence supports my claim that [state claim]. Point out the reasoning that links claim and evidence.*

Record & Reflect:

- Did Chatty Geeps bridge the logic clearly or assume too much?
- Could it explain why this matters historically?

Next Move: Ask Chatty Geeps to find the weakest link in your reasoning chain.

The Counterclaim Challenger

Objective: Learn to anticipate and evaluate alternative interpretations.

Prompt to Try: *Offer a counterargument to my claim about [topic]. Use historical reasoning, not just disagreement.*

Record & Reflect:

- Did the counterargument reveal complexity?
- How could you revise your original claim to address it?

Next Move: Ask Chatty Geeps to list what evidence each side would need to verify its view.

The Big Picture Test

Objective: Connect your claim to larger historical meaning.

Prompt to Try: *Explain why my claim about [topic] matters for understanding broader historical change or continuity.*

Record & Reflect:

- Did Chatty Geeps link your claim to larger patterns or stay narrow?

Next Move: Ask Chatty Geeps to explain how historians might debate its significance.

The Historical Writing Coach

Objective: Scaffold analytical and evidence based-historical essay writing.

Prompt to Try: *Act as a writing coach in history education for a student writing a history essay on [topic]. Suggest an introduction that includes a strong thesis, three evidence-based arguments that cite sources, and a concluding statement.*

Record & Reflect:

- How could the AI-generated example improve student writing?
- Was the thesis clear and effective?
- How can you improve the prompt to develop a stronger output?

Next Move: Write the paper, then ask Chatty Geeps for feedback.

The Intersectional Lens

Objective: Analyze historical texts through an intersectional lens.

Prompt to Try: *Examine how gender, race, and class intersect in [specific historical document, e.g., the Declaration of Sentiments or a speech by Frederick Douglass]. How do these intersections shape the text's message and its historical impact?*

Record & Reflect:

- What intersections did Chatty Geeps (or you) notice most clearly?
- Which dimensions of identity seem emphasized or silenced in the text?
- How do those overlaps change your understanding of the author's purpose or risk?

Next Move: Ask Chatty Geeps to compare this document's intersectional dynamics with another text from a different context or era. What patterns of voice, exclusion, or empowerment persist?

The Source Translator

Objective: Adapt source analysis tasks so multilingual learners can access, question, and use historical evidence without losing disciplinary depth.

Prompt to Try: *Look at this primary source used in my inquiry (e.g., a political cartoon, excerpt, or photograph). How can I scaffold this source for multilingual learners while keeping the task authentic to historical thinking? Generate three supports: one linguistic, one visual, and one conceptual.*

Record & Reflect:

- Which scaffold best balances comprehension and interpretation?
- Did Chatty Geeps suggest tools (glossaries, visuals, translated summaries) that still require reasoning, not rote recall?

Next Move: Ask Chatty Geeps how home languages or peer translation become assets in this task.

The Language Audit

Objective: Examine how colonial language shapes interpretation.

Prompt to Try: *List five words or phrases in [a specific historical source, such as a textbook] that carry colonial bias or hierarchy. Suggest historically accurate but less biased replacements, and explain the difference.*

Record & Reflect:

- How do words naturalize power relations?
- Which replacements clarify rather than sanitize history?

Next Move: Ask Chatty Geeps to compare how translation into another language might alter these power dynamics.

The Global Timeline

Objective: Help students recognize that significant developments occur around the world at the same time as more well-known historical events.

Prompt to Try: *Choose a major historical event, such as the American Civil War (1861–1865). Ask Chatty Geeps to identify major events occurring in other parts of the world during the same period (e.g., the Taiping Rebellion in China, the Unification of Italy, the Second Māori War in New Zealand).*

Record & Reflect:

- What patterns of connection or coincidence do you notice across regions?
- How does seeing multiple events side by side change your sense of significance or causation?

Next Move: Ask Chatty Geeps to help you design a comparative timeline or global map that shows these events happening simultaneously.

The Thread Tangler

Objective: Use Chatty Geeps to trace how seemingly separate historical events were connected through ideas, people, trade, technology, or power, revealing the interdependence of global histories.

Prompt to Try: *Here are two or three events from similar time periods [e.g., the American Civil War and the Taiping Rebellion]. What economic, ideological, or technological connections might link these events, directly or indirectly?*

Record & Reflect:

- Which connections felt solidly evidenced, and which were speculative?
- Did any links reveal power dynamics (colonialism, trade, migration) that change how we interpret local events?

Next Move: Ask Chatty Geeps to help you visualize these connections in a "web of history" diagram, showing how ideas, conflicts, or technologies traveled between regions.

Creating Custom GPTs

Teaching History with Chatty Geeps

The Historian's Apprentice

Objective: Imagine what a GPT could do to strengthen historical thinking.

Prompt to Try: *If I could create a GPT to help students think like historians, what would its main job be (e.g., ask better questions, check evidence, spark curiosity)? List three possible roles.*

Record & Reflect:

- Which role excites you? Which feels risky?

Next Move: Create a name and one-sentence purpose statement for your custom GPT, then ask Chatty Geeps to help design guardrails and instructions for this custom GPT.

The Classroom Companion

Objective: Align your GPT's purpose with real teaching and learning needs.

Prompt to Try: *Describe one classroom challenge (e.g., students struggle to frame questions, cite evidence, or compare sources). How could a GPT scaffold (not solve) this challenge?*

Record & Reflect:

- What balance should exist between guidance and overhelping?

Next Move: Draft the first prompt you'd give students to start using your custom GPT.

The Ethical Guardrail

Objective: Build values and limits into your GPT's design.

Prompt to Try: *List ethical principles your GPT should follow (e.g., transparency, avoiding bias, citing uncertainty, not answering immediately). For each, describe how it would show up in its responses.*

Record & Reflect:

- Which guardrail will be hardest to enforce?

Next Move: Ask Chatty Geeps to critique your list for feasibility and clarity.

The Voice & Tone Curator

Objective: Decide what kind of persona your GPT should adopt.

Prompt to Try: *Describe your GPT's ideal tone and personality. Should it sound like a mentor, peer, coach, historian, or archivist? Why?*

Record & Reflect:

- How does tone affect trust and engagement?

Next Move: Write a short sample reply in that tone, and test it on a peer.

The Prompt Architect

Objective: Build the core instructions for your GPT's behavior.

Prompt to Try: *Create 3–5 guiding rules or sample prompts your GPT will always follow (e.g.,* Ask follow-up questions before answering, Always cite evidence, *or* Use SCIM-C language*).*

Record & Reflect:

- Which rule most strengthens historical reasoning?

Next Move: Reorder your list from most to least essential.

The Source Interpreter

Objective: Design how your GPT handles primary sources.

Prompt to Try: *How should your GPT guide users through analyzing a source? Write a sample exchange using your favorite framework (SCIM-C, RICH, OPTIC, etc.).*

Record & Reflect:

- Did Chatty Geeps model inquiry or shortcut it?

Next Move: Add a safety step in which it reminds users to verify AI-generated interpretations.

The Inquiry Builder

Objective: Create structures for asking and refining questions.

Prompt to Try: *Design a conversation flow. How will your GPT help students move from broad curiosity to a strong, evidence-based inquiry question?*

Record & Reflect:

- Did it encourage reasoning, not guessing?

Next Move: Add a "pause and reflect" step before students search or conclude.

The Feedback Loop

Objective: Build formative feedback into your GPT's responses.

Prompt to Try: *What kind of feedback should your GPT give: coaching, questioning, exemplars, or rubrics? Write an example response to a weak historical claim.*

Record & Reflect:

- Did the tone feel helpful or robotic?

Next Move: Ask Chatty Geeps to annotate why that's a weak response.

The Testing Table

Objective: Prototype and stress-test your GPT.

Prompt to Try: *Ask your GPT to do something difficult (e.g., interpret bias, assess evidence, spot gaps). Note what it does well, what it misreads, and how you could refine the instructions.*

Record & Reflect:

- Where did it succeed because of your setup?
- Where did it falter due to unclear guidance?

Next Move: Revise your instructions or narrow your conversation starters accordingly.

Assessment

Teaching History with Chatty Geeps

The Technocurious Observer: Rethinking What Counts

Objective: Reimagine what assessment captures beyond correctness.

Prompt to Try: *List three ways GenAI tools could help me see evidence of student thinking that a traditional quiz would miss. Include ways that privilege process, creativity, or reflection.*

Record & Reflect:

- Did Chatty Geeps emphasize products or processes?
- What would it take to value curiosity itself?

Next Move: Ask Chatty Geeps to design a quick formative check that measures growth in curiosity.

The Feedback Loop: Learning from the Machine's Mirror

Objective: Use Chatty Geeps's feedback to deepen reflection instead of replace judgment.

Prompt to Try: *Review the following paragraph. Identify strengths and one area for growth. Explain your reasoning using accessible, student-friendly language.*

Record & Reflect:

- Was the feedback actionable? Humane? Overly generic? Specific to the right discipline?
- What does this reveal about the limits of Chatty Geeps in formative assessment?

Next Move: Ask Chatty Geeps to model a student response to the feedback and judge if it leads to improvement.

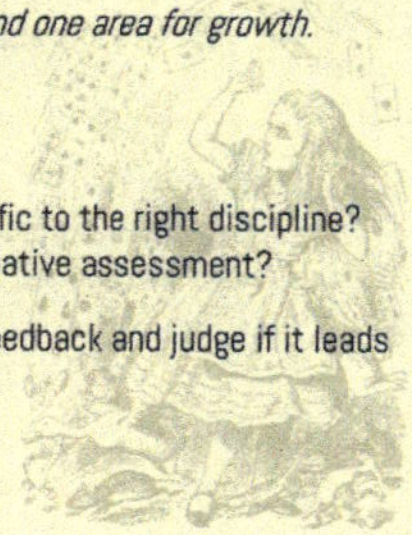

The Evidence Collector: Tracing Growth over Time

Objective: Use Chatty Geeps to help visualize progression in historical thinking.

Prompt to Try: *Compare two of my analyses (early vs. more recent). Identify evidence of progression in causal reasoning and use of evidence.*

Record & Reflect:

- Did Chatty Geeps focus on depth, vocabulary, or accuracy?
- How might you represent your growth visually?

Next Move: Ask Chatty Geeps to propose sentence stems you could use to self-identify growth (e.g., "I used to think..., now I think...").

The Mapmaker's Rubric: Designing Fitness for Purpose

Objective: Codesign rubrics that measure what you truly value.

Prompt to Try: *Based on these assignment instructions provided by the teacher, create a rubric for assessing student work that emphasizes interpretation, evidence use, and reflection rather than correctness.*

Record & Reflect:

- Did Chatty Geeps's categories match the disciplinary aims of the assignment?
- Which descriptors felt too rigid?

Next Move: Students rewrite one criterion in their own words to build shared understanding.

The Formative Alchemist: Turning Tasks into Learning Moments

Objective: Transform existing tasks into opportunities for metacognition.

Prompt to Try: *Consider the following assessment task. Suggest one AI-supported variation that turns it into a formative learning opportunity instead of a summative test.*

Record & Reflect:

- Did Chatty Geeps retain disciplinary rigor while adding reflection?
- What might students learn about their own processes?

Next Move: Pilot the new version with a small group, and note differences in engagement.

The Self-Regulator: Students as Co-Assessors

Objective: Use Chatty Geeps to help students monitor and plan their learning.

Prompt to Try: *Create a self-check prompt students could use with AI to assess how well they applied [a specific scaffolding strategy; e.g., OPTIC, SCIM-C, RICH].*

Record & Reflect:

- Does the Chatty Geeps model encourage honest self-assessment or superficial praise?
- How might you build this into regular classroom routines?

Next Move: Ask Chatty Geeps to design a reflection journal format for ongoing metacognitive tracking.

The Observatory: Seeing Patterns, Not Just Scores

Objective: Use Chatty Geeps to expose classwide patterns for responsive teaching.

Prompt to Try: *Summarize key strengths and needs across these ten student reflections.*[1] *Identify one emerging misconception about causation or evidence use.*

Record & Reflect:

- Did Chatty Geeps capture nuance or oversimplify trends?
- How could you verify its pattern recognition?

Next Move: Ask Chatty Geeps to suggest a mini-lesson addressing the pattern.

1. Note: If you upload student work into Chatty Geeps, be sure it is de-identified first.

The Learning Mirror: What Did I Actually Learn?

Objective: Reflect on what your thinking shows.

Prompt to Try: *Help me describe what I've learned so far about [topic]. Ask me questions that reveal what has changed in how I think about history.*

Record & Reflect:

- What surprised you about your own answers?
- Did Chatty Geeps help you notice growth or repetition?

Next Move: Ask Chatty Geeps to help you turn your reflections into a visual map of your learning journey.

The Feedback Translator: Making Sense of Comments

Objective: Learn to interpret teacher feedback in your own words.

Prompt to Try: *Review the following feedback I received. Help me restate what it means and what steps I could take to address the feedback.*

Record & Reflect:

- How does this prompt help support my learning?
- Which suggestion felt most doable?

Next Move: Ask Chatty Geeps to show you what action steps to take, then decide which ones you want to keep or change.

The Revision Lab: Using Chatty Geeps as a Practice Partner

Objective: Use Chatty Geeps's feedback to refine, not replace, your ideas.

Prompt to Try: *Read this paragraph, and give me feedback focused on clarity, evidence, and reasoning. Don't rewrite it; ask me questions to help me think.*

Record & Reflect:

- Did Chatty Geeps's questions help you clarify your argument?
- What ideas did you keep, change, or reject?

Next Move: Ask Chatty Geeps to show what a revised version might sound like, then compare yours to its example. You can submit both copies and annotate what suggestions you kept and what you did differently.

The Source Detective: How Am I Using Evidence?

Objective: Reflect on your use of sources and historical habits of mind.

Prompt to Try: *Ask me three questions to check how well I used sources in my project on [topic]. Then help me think about what stronger sourcework would look like.*

Record & Reflect:

- Did you quote, summarize, or interpret most often?
- What could make your next source analysis more convincing?

Next Move: Ask Chatty Geeps to help you write a one-sentence goal for your next research task.

The Reflective Historian

Objective: Reflect on how inquiry shapes understanding.

Prompt to Try: *Ask me three questions that help me evaluate my own inquiry: what I learned, what I'm still unsure about, and how my thinking changed.*

Record & Reflect:

- Did Chatty Geeps prompt genuine reflection or checklist answers?

Next Move: Write a one-sentence summary: "My understanding changed when I realized..."

The Multilingual Historian

Objective: Design assessments that help multilingual learners show what they know through multiple modes of expression.

Prompt to Try: *How can I adapt a rubric or feedback system for this inquiry so that it recognizes historical thinking, use of evidence, and progress for multilingual learners? Provide an example of a rubric criterion rewritten for inclusivity.*

Record & Reflect:

- Did Chatty Geeps balance disciplinary rigor with sensitivity to language learning?
- What kinds of evidence—e.g., oral explanations, visuals, translated notes—could demonstrate understanding even when writing is limited?

Next Move: Ask Chatty Geeps to model a short piece of narrative feedback for a multilingual student that highlights strengths in reasoning and suggests one specific linguistic stretch goal.

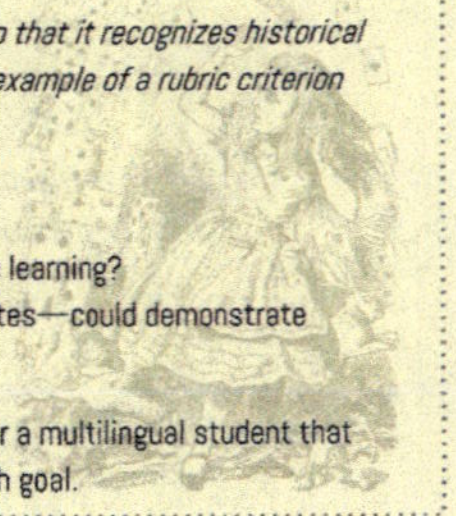

Prompting and Asking Questions

Teaching History with Chatty Geeps

The Historian's Key: Opening the Question

Objective: Learn to frame prompts that begin inquiry, not end it.

Prompt to Try: *Help me write three different types of historical questions about [topic]: one focusing on causation, one on continuity and change, and one on significance.*

Record & Reflect:

- Which question type led to richer exploration?
- Did Chatty Geeps favor fact recall or interpretation?

Next Move: Ask Chatty Geeps to reframe one question as a mystery or puzzle for students to solve.

The Inquiry Architect

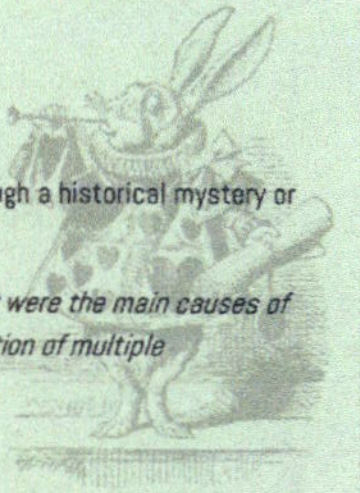

Objective: Use Chatty Geeps to design a scaffolded investigation that guides learners through a historical mystery or debate with structure, evidence, and reflection.

Prompt to Try: *Create a step-by-step investigation plan for a historical question (e.g., What were the main causes of the fall of the Roman Empire?). The plan should include primary source analysis, consideration of multiple perspectives, checkpoints for student reflection, and a final evidence-based conclusion.*

Record & Reflect:

- Did Chatty Geeps's outline promote inquiry rather than give away answers?
- Where could scaffolds better support diverse learners or multilingual students?
- Which steps genuinely develop historical thinking (sourcing, corroboration, argumentation)?

Next Move: Ask Chatty Geeps to generate three variations of the same investigation: one for emergent or developing students, one for proficient, and one for sophisticated learners. Compare how each balances structure, autonomy, and disciplinary depth.

The Context Builder: Setting the Stage

Objective: Explore how context shapes Chatty Geeps's responses.

Prompt to Try: *Explain [event] as if you were teaching it to students who know nothing about the time period. Include details about place, people, and historical conditions.*

Record & Reflect:

- Did Chatty Geeps include essential context or skip too quickly to summary?

Next Move: Ask Chatty Geeps what background information it assumed; then check that assumption.

The Framing Experiment: Changing the Lens

Objective: Test how different framing changes historical interpretation.

Prompt to Try: *Describe [event] through the following lenses: social, political, economic, religious, and military. What changes in emphasis or meaning?*

Record & Reflect:

- Which lens created the most surprising interpretation?
- Did Chatty Geeps's tone or vocabulary shift noticeably?

Next Move: Ask Chatty Geeps to compare which lens a historian from the 1920s might have used and why.

The Follow-Up Trick: Prompt Chaining

Objective: Practice follow-up questioning to deepen understanding.

Prompt to Try: *Summarize [event]. Now explain why historians disagree about its causes. Show me two different interpretations and the evidence for each.*

Record & Reflect:

- How did layering questions change the quality of Chatty Geeps's reasoning?

Next Move: Design your own three-step "prompt chain" for a history topic you teach.

The Source Whisperer: Working with Evidence

Objective: Learn to prompt Chatty Geeps to think with, not for, the historian.

Prompt to Try: *Review the following short excerpt from a primary source. Ask me three questions that help me interpret it, and then summarize what I've said.*

Record & Reflect:

- Did Chatty Geeps ask authentic interpretive questions?
- What did its questions miss about voice, motive, or context?

Next Move: Reverse roles. You write the questions and have Chatty Geeps answer them.

The Detective's Doubt: Testing Chatty Geeps's Confidence

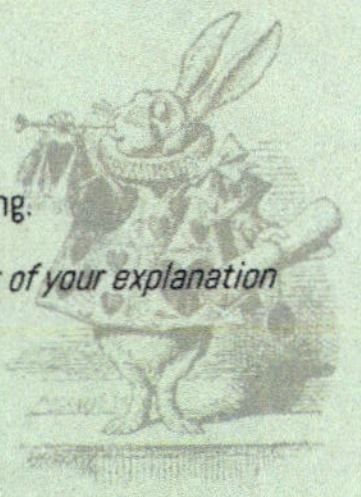

Objective: Encourage skepticism by interrogating Chatty Geeps's reasoning.

Prompt to Try: *Explain why the [event] happened. Then tell me which part of your explanation might be uncertain or debated.*

Record & Reflect:

- Did Chatty Geeps acknowledge limits or hedge its answers?

Next Move: Ask it to cite possible historians who might disagree and why.

The Scaffold Maker: Prompting for Historical Thinking

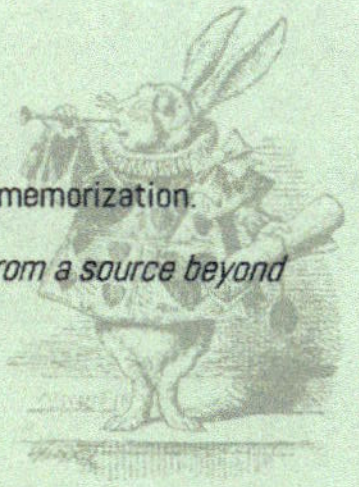

Objective: Design prompts that help students demonstrate thinking, not memorization.

Prompt to Try: *Create a prompt that requires students to infer meaning from a source beyond just summarizing it.*

Record & Reflect:

- Did Chatty Geeps's prompt elicit interpretation or recall?

Next Move: Ask Chatty Geeps to adjust the same prompt for four progression levels (emergent, developing, proficient, sophisticated).

The Stress Test: Fact-Checking the Machine

Objective: Use prompting to verify Chatty Geeps's accuracy.

Prompt to Try: *Give me three claims about [event]. Then tell me how I could verify each one using historical sources.*

Record & Reflect:

- Which sources did it recommend?
- Did it prioritize credibility or convenience?

Next Move: Ask Chatty Geeps to rank those sources by reliability and justify its order.

The Prompt Autopsy: Reflecting on What Worked

Objective: Build metacognition about prompting technique.

Prompt to Try: *Review the following prompt I used. Analyze what made it effective or unclear. Suggest a revised version for deeper historical thinking.*

Record & Reflect:

- Did Chatty Geeps notice ambiguity or lack of context?
- What rules of good prompting can you infer from this exercise?

Next Move: Collect three of your "best prompts," and explain why they worked.

The Perspective Taker: Chatty Geeps as Dialogue Partner

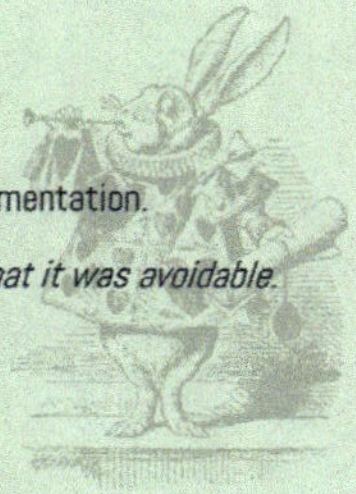

Objective: Treat Chatty Geeps as a conversation partner to practice argumentation.

Prompt to Try: *Argue that [historical event] was inevitable. Then argue that it was avoidable. Help me compare the strength of both arguments.*

Record & Reflect:

- Which argument seemed better supported by evidence?
- Did Chatty Geeps show awareness of historical contingency?

Next Move: Ask Chatty Geeps to summarize what it learned from your counterarguments.

The Compass Question: Finding What's Worth Asking

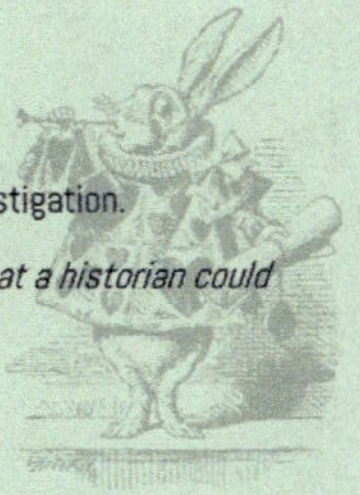

Objective: Learn to frame compelling historical questions that spark investigation.

Prompt to Try: *Help me write three compelling questions about [topic] that a historian could actually investigate with evidence. Explain why each question matters.*

Record & Reflect:

- Which question pulls you toward curiosity rather than quick facts?
- Did Chatty Geeps understand what makes a question historical?

Next Move: Ask Chatty Geeps to turn your strongest question into a classroom inquiry prompt.

The Supporting Crew: Breaking Big Questions into Small Ones

Objective: Practice developing supporting questions that guide inquiry.

Prompt to Try: *Consider the following compelling question and suggest five smaller supporting questions that help answer it.*

Record & Reflect:

- Which supporting questions require evidence, not opinion?
- Where do these questions help you move from curiosity to investigation?

Next Move: Ask Chatty Geeps to match each supporting question with possible types of sources.

The Inquiry Mapmaker

Objective: Visualize how questions, evidence, and claims interconnect.

Prompt to Try: *Help me make a short outline that connects my compelling question, supporting questions, evidence, and final claim.*

Record & Reflect:

- What links felt strongest or weakest?
- Where do you still need evidence?

Next Move: Sketch this as an inquiry diagram or timeline.

The Inquiry Bridge

Objective: Evaluate and adapt inquiry questions so English learners and students with limited background knowledge can engage meaningfully with historical inquiry.

Prompt to Try: *Review this compelling question: Was the New Deal a success? Suggest scaffolds or modifications that would help multilingual learners participate fully in exploring this question and its supporting sources.*

Record & Reflect:

- What visuals, sentence frames, or pre-inquiry activities could help learners access key ideas before tackling texts?
- How could students draw on their own linguistic and cultural experiences to define success in ways that enrich the inquiry?

Next Move: Ask Chatty Geeps to generate three accessible versions of the same IDM question: one that simplifies language, one that builds context, and one that invites cross-cultural comparison. Discuss which version best preserves historical rigor while widening participation.

The Inquiry Evaluator

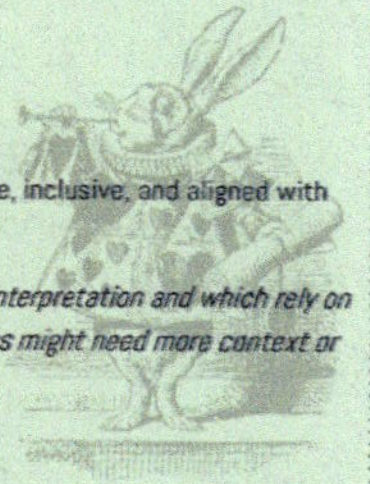

Objective: Critically assess and refine AI-designed inquiry plans to ensure they are accurate, inclusive, and aligned with disciplinary goals in history education.

Prompt to Try: *Looking at this inquiry lesson plan, tell me which parts of this plan rely on interpretation and which rely on fact, how it handle bias or missing perspectives, and where multilingual learners or novices might need more context or vocabulary support.*

Record & Reflect:

- Which parts of the plan encourage genuine inquiry, and which lean toward rote procedure?
- Did Chatty Geeps's feedback consider diverse voices and ethical awareness?
- How can you make the plan more transparent about uncertainty or contested evidence?

Next Move: Revise one stage of the plan yourself: add or reframe a question, swap in a more inclusive source, or adjust scaffolds for accessibility. Then ask Chatty Geeps to explain how your revision changes the learning outcomes.

Prompting as an Archival Tool

The following prompts can be used to help define search terms when searching the Library of Congress or the National Archives.

Teaching History with Chatty Geeps

The Keyword Archaeologist

Objective: Practice searching databases using historically aware language.

Prompt to Try: *I'm researching [topic]. List ten historical keywords or phrases I might use in an archive search from different time periods and perspectives.*

Record & Reflect:

- Did Chatty Geeps include outdated or period-specific terms (e.g., "Negro" vs. "African American")?
- What does this reveal about how language shapes access to sources?

Next Move: Ask Chatty Geeps to explain which search terms might be harmful or biased and what modern equivalents exist.

The Date Miner

Objective: Learn to filter searches by relevant time frames.

Prompt to Try: *I'm studying reactions to [event]. Suggest a range of years before and after the event when primary sources might still reflect its impact.*

Record & Reflect:

- Did Chatty Geeps over-restrict or over-broaden the timeline?

Next Move: Ask Chatty Geeps to justify why those date ranges matter historically (e.g., for their continuity, aftermath, memory).

The Synonym Sleuth

Objective: Explore how word choice limits or expands search results.

Prompt to Try: *Generate a list of synonyms, spelling variations, or historical terms for [topic] that might appear in newspapers or letters.*

Record & Reflect:

- Which terms feel anachronistic or inaccurate?
- How could this list improve your database search strategy?

Next Move: Feed these variations into a real database (e.g., Chronicling America), and compare hits between search terms.

The Metadata Detective

Objective: Understand how metadata (tags, categories) organizes information.

Prompt to Try: *Explain what metadata fields I should look for in an archive entry about [topic] (e.g., date, creator, location, format). Why do they matter?*

Record & Reflect:

- Which fields most affect the meaning of a source?
- What metadata might mislead if misinterpreted?

Next Move: Ask Chatty Geeps to design a short "metadata checklist" for evaluating sources.

The Catalog Whisperer

Objective: Practice writing queries that fit database logic.

Prompt to Try: *Help me write a search string that finds newspaper articles about [topic] between [year range] in [region]. Show both a simple and an advanced version.*

Record & Reflect:

- Did the advanced version use operators (AND, OR, NOT, quotes)?
- Which version balanced precision and discovery best?

Next Move: Ask Chatty Geeps to explain the logic behind each operator using your example.

The Query Cartographer

Objective: Map how different search strategies shape what you find.

Prompt to Try: *Give me three different search approaches to study [topic]: one broad, one ballistic, and one comparative. Explain what each might uncover.*

Record & Reflect:

- Which search strategy feels most promising?
- How do you balance specificity with openness?

Next Move: Try the searches, diagramming how your "information map" changes.

The OCR Detector

Objective: Explore how OCR (optical character recognition) technology shapes what historians can and cannot find in digitized archives.

Prompt to Try: *Explain how OCR works and why it often misreads certain letters or fonts. Then compare one clearly scanned article with one full of OCR errors. What kinds of information are lost or distorted?*

Record & Reflect:

- What types of words or characters are most often misread (e.g., long s, ligatures, faded print)?
- How do OCR limitations influence whose stories remain visible in digital archives?
- What happens when errors cluster around specific groups, regions, or topics?

Next Move: Ask Chatty Geeps to help you design a short classroom demonstration in which you show an original newspaper image alongside its OCR text and have students highlight the differences.

The OCR Error Assistant

Objective: Develop historically informed search strategies that account for linguistic change, variant spellings, and shifting place names.

Prompt to Try: *Look at this search term: [name a person, place, or event]. What older spellings, titles, or regional names might have been used during this period? What alternative names could yield additional results?*

Record & Reflect:

- Which historical or linguistic variations revealed new primary sources?
- How does awareness of evolving vocabulary expand the inclusivity of your search?
- What assumptions about language or geography did Chatty Geeps have to correct?

Next Move: Ask Chatty Geeps to build a "historical keyword evolution chart" showing how one term (e.g., "immigration," "labor," "citizenship") changed from the nineteenth to twentieth century.

The Search Planner

Objective: Learn to define historical context, scope, and purpose to guide smarter inquiry.

Prompt to Try: *Help me plan my search context. Please provide a short timeline of key dates or phrases related to my topic; a list of regions, organizations, or newspapers most likely to cover it; terms or titles that contemporaries might have used to describe it; and related social, political, or global events that may shape coverage.*

Record & Reflect:

- How did building context before searching change the quality or relevance of your results?
- Did background suggestions reveal perspectives you might have missed (e.g., local or international contexts)?

Next Move: Ask Chatty Geeps to turn your context notes into a "search plan": a short, tiered guide listing what to look for first, second, and third.

The Term Tracker

Objective: Recognize that place names, colonial labels, and spelling conventions can change, affecting what historians can discover in digital archives.

Prompt to Try: *We are looking at a specific region or city (e.g., Istanbul, Mumbai, Zimbabwe). What historical or colonial names has this place been known by? How have spelling or transliteration conventions changed across time and language? Which versions appear in nineteenth- and early twentieth-century newspapers or records?*

Record & Reflect:

- Which name produced the most diverse results?
- How do naming conventions reflect political, cultural, or colonial power?

Next Move: Ask Chatty Geeps to build a "name evolution table" that lists each variant with dates of use and linguistic context, then use it to rerun your searches.

Using the Primary Sources in This Book

This set of cards works a little differently than the rest. We believe that, if you want to incorporate GenAI in your classroom, you're going to need to get a little messy first. We created these curiosity cards to give YOU an opportunity to play around Chatty Geeps:– to "go live" and experience what we mean when we say technocuriosity requires an amphibious lens.

Here's how we recommend using them. First, do an analysis of the provided source without Chatty Geeps. You can use the prompt questions we provide or your favorite analysis strategy. Then follow the instructions on the card to prompt Chatty Geeps and reflect. How do these activities help you think about what it means to build historical thinking skills using GenAI as a scaffold?

Teaching History with Chatty Geeps

The Mediaeval Manor and Its Afterlives

Objective: To question how early twentieth century historians imagined the "order" of medieval society, and to explore how ideas of hierarchy and land use in such maps echo later systems of classification, empire, and data control.

Prompt to Try: *Look at this 1923 map titled* Plan of a Mediaeval Manor *by historian William R. Shepherd. What argument about the Middle Ages does this map make? What story of order, labor, or hierarchy does it suggest? Explain how the map's neat fields and labels might reflect the worldview of early twentieth-century historians rather than actual medieval life. How might ideas of "organization" and "improvement" in this map echo later systems like colonial mapping or modern data visualization? Suggest one change you would make to this map to show a more inclusive or complex vision of community.*

Record & Reflect:

- How does the map's geometry (neat boundaries, labeled fields) express a particular idea of order and control?
- What assumptions about labor, gender, or class might be hidden behind its tidy layout?
- Why might a 1923 historian find this vision comforting or educational?
- What happens when we read the map not as a description but as an argument about civilization?

Next Move: Ask yourself what this exchange reveal about how Chatty Geeps "knows" history and how our tools inherit older patterns of inclusion and exclusion.

Innocence, Nature, and the Moralizing of Curiosity

Objective: To explore how Victorian culture imagined childhood curiosity as both pure and in need of control, and to question how those same ideas about "innocence," "nature," and "improvement" shape the stories we build today.

Prompt to Try: *Examine Linley Sambourne's 1895 illustration from* The Water-Babies *by Charles Kingsley. Describe how this image imagines childhood curiosity. What is the child doing, feeling, or discovering? Explain what moral or scientific lesson a Victorian reader might see in this scene. Why might a culture that prized innocence also fear curiosity? Rewrite the caption to reflect a 2025 understanding of curiosity that values wonder without moral judgment.*

Record & Reflect:

- How does the image turn curiosity into something to admire or to rein in?
- Who is allowed to be curious here? Who or what is acted upon instead?
- How do these ideas echo in today's narratives of childhood, learning, or even AI "innocence"?

Next Move: Pair the historical image with a modern photograph, meme, or AI-generated image, and ask Chatty Geeps to analyze both. How do style and symbolism shape truth-telling?

Satire, Sympathy, and the Stories We Believe

Objective: To investigate how Victorian media used satire to shape public opinion and moral judgment, and to consider how similar dynamics (credibility, sympathy, and spectacle) reappear in today's digital and algorithmic cultures.

Prompt to Try: *Examine the 1872 cartoon* The Beggar's Petition *from* Judy; or the London Serio-Comic Journal. *Explain how the artist uses exaggeration, irony, and labeling to shape how viewers judge the beggar. What social or moral assumptions about poverty, truth, or respectability are embedded in this satire? How might this cartoon have influenced Victorian readers' opinions about the Tichborne case or about charity in general?*

Record & Reflect:

- What feelings does the image try to provoke? Pity, ridicule, outrage?
- How does visual satire work as both journalism and entertainment?
- Why might societies use humor to enforce moral boundaries or social order?
- In what ways do digital media today still use humor, bias, or repetition to make some people seem trustworthy and others not?"

Next Move: Experiment with rewording part of the prompt to shift Chatty Geeps's stance by providing a specific perspective (*You are a historian* or *You are a journalist*). How does the model's reasoning or bias change when you alter its perspective?

Power, Maps, and Meaning

Objective: To investigate how early modern mapping combined discovery, imagination, and power, and to compare how both historical and digital maps shape who and what gets seen, named, and valued.

Prompt to Try: *Examine the 1507 Waldseemüller Map, the first to label the new continents as America. Explain how this map blends knowledge and imagination. What does it get "right," and what does it invent? Why might the act of naming a continent after Amerigo Vespucci be historically and politically significant? Discuss how maps are never neutral. What power does naming or outlining territory give to the mapmaker?*

Record & Reflect:

- What stories about the world does this map tell?
- How does naming make knowledge feel official or permanent?
- How might modern maps still reflect unequal access to power, technology, or representation?

Next Move: Ask Chatty Geeps to describe one continuity and one change between the world of the source and today. What patterns of thought or representation persist?

Evidence, Wonder, and the Medieval Mind

Objective: To explore how medieval thinkers blended imagination and observation to make sense of the natural world, and to reflect on how modern technologies still create meaning from limited or uncertain evidence.

Prompt to Try: *You are a modern historian examining this image of a dragon attacking an elephant from the Aberdeen Bestiary (c. 1200 AD). Explain what this illustration suggests about how medieval people imagined creatures they had never seen. How did evidence, imagination, and morality combine to create "knowledge" in this period? Compare this process to how AI systems today generate images or information about things we don't fully understand. What do both the bestiary and modern AI reveal about human curiosity and the need to make sense of the unknown?*

Record & Reflect:

- How does the image turn curiosity into a moral story about good and evil?
- Why might a society trust images that mix truth and fantasy?
- How do we still do that today with data, media, or algorithms?

Next Move: Ask Chatty Geeps to "fill in" missing details or context from the source, then fact-check its claims with real evidence. Where does Chatty Geeps speculate, and how can history educators respond critically?

Ada Lovelace and the History We Write

Objective: To examine how Ada Lovelace's authorship was recorded, minimized, and later rediscovered, revealing how gender shaped who was recognized as a "knower" in the making of modern technology.

Prompt to Try: *You are a historian analyzing Ada Lovelace's 1843* Note G, *often called the world's first computer program. Explain what made Lovelace's contribution significant and why it was long overlooked. How did Victorian ideas about gender and intellect shape who was credited for innovation in science and mathematics? Identify phrases or structures in the diagram that show her blending logic with imagination. What does that reveal about how she understood "thinking machines"?*

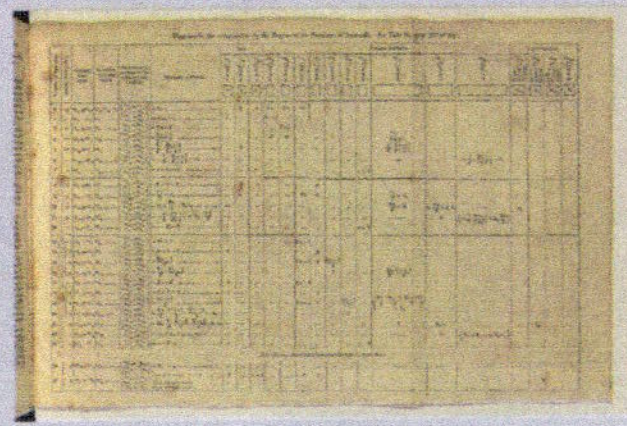

Record & Reflect:

- What barriers did Lovelace face that her male collaborators did not?
- How does the tone or format of *Note G* show both brilliance and constraint?
- How does her vision challenge the idea that creativity and logic are opposites?

Next Move: Reflect: What does the model emphasize? Genius, collaboration, erasure, or myth? How might its framing echo or challenge Victorian narratives about gender and intellect?

Mary Anning and the Shape of Discovery

Objective: To analyze how Mary Anning's fieldwork and fossil illustrations challenged nineteenth-century ideas about who could be a "scientist," and to consider how evidence, skill, and credibility are still shaped by visibility and value systems today.

Prompt: *You are a historian studying Mary Anning's 1824 fossil drawing. Describe what makes this document both scientific and artistic. How did Mary Anning's position as a self-taught woman in early nineteenth-century England affect how her discoveries were credited or recorded? What does this image reveal about how scientific evidence is constructed, interpreted, and communicated? Compare how knowledge was validated in Anning's time to how it's validated today in AI-generated or data-driven research.*

Record & Reflect:

- How might gender, class, and education have shaped Anning's access to recognition?
- Why might scientific knowledge appear objective but still reflect social hierarchies?
- How do modern systems (like AI or academia) echo or resist those hierarchies?

Next Move: Try asking Chatty Geeps to describe this same fossil drawing "as if discovered by a famous male scientist," then compare tone and authority. What does this reveal about how bias circulates through data and description?

Belonging and Memory at Manzanar

Objective: To analyze how Adams's photograph frames community and confinement during Japanese American incarceration in WWII, and to reflect on how images shape collective belonging and national memory.

Prompt to Try: *You are a photo historian analyzing Ansel Adams's 1943 photograph of a baseball game at the Manzanar Relocation Center. Describe how Adams uses composition, light, and perspective to create a sense of normalcy or dignity. Why might he have chosen to frame the camp through recreation rather than restriction? How does this photograph both reveal and obscure the injustice of incarceration?*

Record & Reflect:

- What emotions or narratives does the image invite and which does it avoid?
- How might beauty in an image of injustice complicate remembrance?
- In what ways do modern technologies continue this tension between visibility and control?

Next Move: Reflect on continuity and change. How do modern technologies like social media, surveillance, or AI image generators shape who and what is seen as part of the national story?

Media, Emotion, and the Power of Exaggeration

Objective: To examine how nineteenth-century artists used visual exaggeration and fantasy to shape public emotion and ideology, and to compare those tactics with the ways digital images and memes influence belief today.

Prompt: *You are a critical viewer analyzing the 1853 lithograph* A Dream Caused by the Perusal of Uncle Tom's Cabin. *What emotions is this image designed to stir in its viewers? Which groups or ideas are portrayed as heroic, ridiculous, or dangerous? How does exaggeration turn opinion into ideology?*

Record & Reflect:

- What visual cues reveal the artist's bias or target audience?
- How might humor or horror make false messages feel persuasive?
- Whose fears are amplified here, and whose voices are silenced?
- Where do you see similar emotional strategies in today's online media?

Next Move: Compare this nineteenth-century use of visual fear to modern digital misinformation. How do memes, deepfakes, or viral images still use distortion to shape belief?

News, Emotion, and the Road to War

Objective: To analyze how early twentieth-century newspapers framed the causes and emotions of a global crisis, and to explore how tone and imagery made the coming war feel both shocking and inevitable.

Prompt to Try: *You are a historian analyzing this 1914* New York American *newspaper page about the assassination of Archduke Franz Ferdinand. Describe the tone and mood of the coverage. What emotions does the newspaper invite readers to feel? How do headlines, portraits, and quotations shape the sense that war might be unavoidable? What connections or assumptions about blame, empire, and fate does this page encourage? If you were writing a headline about the same event, how might you frame it differently to emphasize contingency instead of inevitability?*

Record & Reflect:

- What words or images suggest crisis, destiny, or tragedy?
- How do media narratives transform single events into global turning points?
- Where can we see causation being constructed rather than described?
- What makes certain explanations of events "stick" in public memory?

Next Move: Ask Chatty Geeps to rewrite one of the included articles using three different editorial voices: a pacifist, a nationalist, and a modern journalist trained in data-driven neutrality. Compare how each version reframes causation and consequence.

Alice and Victorian Satire

Objective: To explore how *Punch* used Lewis Carroll's Alice to satirize the confusion, contradictions, and self-importance of late Victorian government, and to consider how familiar cultural symbols become tools for social commentary.

Prompt to Try: *You are a Victorian critic explaining why* Alice in Wonderland *became such a popular metaphor for politics and bureaucracy in the late 1800s. How does this* Punch *cartoon,* Alice in Bumbleland, *transform Carroll's imaginative world into a satire about government or social order? What clues in the image (costume, dialogue, or expressions) help readers decode its targets? Why might Victorians have found the idea of nonsense a useful way to critique authority? What happens when political systems become so complex that they start to resemble the absurd worlds of fiction?*

Record & Reflect:

- What elements of the Alice story survive here, and what's been twisted for humor or critique?
- How does seeing Alice in a political cartoon change your sense of her character from innocent explorer to political observer?
- What does this say about how stories travel between art, politics, and culture?
- How might satire help societies question systems that claim to be logical?

Next Move: Invite Chatty Geeps into a "cross-century conversation." Ask it to generate two short responses to the cartoon: one as a Victorian editor at *Punch* explaining its message to readers and one as a modern journalist reflecting on political satire today.

Power, Perspective, and the Fall of Bismarck

Objective: To explore how *Punch* used nautical metaphor and visual perspective to comment on the resignation of German Chancellor Otto von Bismarck, and to consider how images can make complex political shifts feel natural, inevitable, or tragic.

Prompt to Try: *You are a Victorian political analyst writing in 1890. Describe what is happening in the* Punch *cartoon,* Dropping the Pilot. *Who are the two figures, and what moment does this represent? How does the artist use setting, posture, and symbolism to shape readers' sympathy? What tone or judgment about leadership and change does this image suggest? Why might Tenniel choose a ship as the setting? What does that choice imply about control, direction, and national destiny?*

Record & Reflect:

- How does Tenniel's viewpoint guide the audience's emotional response?
- What makes the image both respectful of Bismarck and critical of Wilhelm II?
- How can a single, silent image communicate political transition more effectively than words?

Next Move: Invite Chatty Geeps into a visual-translation experiment. Ask it to reinterpret *Dropping the Pilot* using a modern political analogy or corporate metaphor (e.g., a CEO stepping down, an algorithm being retired, a leader replaced by automation).

Race, Representation, and the Politics of Respectability

Objective: To analyze how early twentieth-century media used imagery and tone to police racial and social boundaries, and to consider whose perspectives were amplified or erased in public debates about equality and leadership.

Prompt to Try: *You are a historian analyzing this 1901* Times *cartoon published after President Theodore Roosevelt invited Booker T. Washington to dinner at the White House. Describe what is happening in the image and who the figures represent. How does the cartoon's composition, body language, and caption reveal the artist's racial and political perspective? What emotions or fears is this cartoon trying to provoke in its readers?*

Record & Reflect:

- What does this cartoon suggest about who was "allowed" to belong in political life at the time?
- How does tone (mockery, ridicule, or fear) work to maintain social hierarchy?
- What risks did Washington and Roosevelt each face for crossing racial boundaries in a public act?
- How does seeing this through the lens of 1901 media help us understand the persistence of representational power?

Next Move: Reflect. How can historians and teachers confront harmful imagery without repeating its harm?

Democracy in Wonderland

Objective: To analyze how contemporary political cartoons use familiar cultural metaphors like *Alice in Wonderland* to comment on power, democracy, and the fragility of political norms.

Prompt to Try: *You are a cultural historian examining Rob Rogers's 2020 cartoon* Trump's Willing Executioner, *which reimagines* Alice in Wonderland *to comment on American democracy. Explain why Alice remains such a powerful metaphor for disorder, absurdity, and power in politics. How does the cartoon use characters (the Queen of Hearts, Alice, the executioner) to symbolize attitudes toward democracy, authority, and obedience? What emotions (fear, irony, defiance) does the cartoon try to evoke and in which audiences? Why might this image be historically significant in the political climate of 2020?*

Record & Reflect:

- What parts of Carroll's original story are being borrowed, and which are reimagined?
- How does labeling the girl *DEMOCRACY* change the reader's emotional response?
- What does this cartoon suggest about the state of political debate and trust in institutions?
- How might future historians interpret this cartoon as evidence of democratic tension?

Next Move: Run the same historical question with two different prompt wordings (one emotional, one factual). Compare how Chatty Geeps's tone and evidence shift. How does asking differently change what you learn?

Kew Gardens as a Living Archive

Objective: To analyze how Victorian Britain used botanical gardens like Kew as sites of imperial science, collection, and control, and to consider how those same logics of cataloguing and display are echoed in modern data systems and algorithmic "archives."

Prompt to Try: *You are a modern historian analyzing this 1850s engraving of the Great Palm House at Kew Gardens. Describe what this image communicates about science, order, and empire in Victorian Britain. How does the design of Kew (its glasshouse, gardens, and promenading visitors) symbolize control over nature and the world beyond Britain? What does it mean to call Kew an "imperial archive of living things"? Compare this system of collection and classification to modern digital databases or AI datasets. How do both organize the world to make it seem knowable?*

Record & Reflect:

- Who is visible in this scene, and who is not?
- What kinds of knowledge are being displayed, and who controls access to them?
- How does beauty and order mask power or exploitation?
- In what ways do gardens, libraries, and databases all serve as mirrors of cultural values?

Next Move: Guide Chatty Geeps in a reflection on systems of order. Ask it to explain what Kew Gardens and a modern AI training dataset have in common. Then challenge it to identify what kinds of things each system cannot or does not collect.

Spirit Photography and the Edges of Evidence

Objective: To investigate how Victorian spirit photography blurred boundaries between science, faith, and imagination, and to reflect on how new technologies create both wonder and skepticism about what is "real."

Prompt to Try: *You are a skeptical observer analyzing this 1877 spirit photograph titled* The Orphans at Their Mother's Grave. *Describe what you see. What story does the image want viewers to believe? How does the faint figure of the mother change the emotional meaning of the photograph? Why might Victorians have trusted photography as proof of spiritual or unseen worlds? Compare this to modern technologies (like deepfakes, AI-generated images, or virtual memorials). How do new tools still mix belief, evidence, and deception?*

Record & Reflect:

- What kinds of evidence does this image provide or pretend to provide?
- How do emotion and technology reinforce one another here?
- How do we decide what's authentic when new media make the impossible visible?

Next Move: Guide Chatty Geeps into an inquiry about belief and verification: Ask it to identify how the logic of spirit photography resembles modern uses of AI-generated imagery or misinformation.

War, Memory, and the Camera's Gaze

Objective: To examine how war photography shapes collective memory, emotion, and national identity, and to consider how iconic images carry the burden of historical representation.

Prompt to Try: *You are a media critic analyzing the famous 1944 photograph* Into the Jaws of Death. *What emotions or ideas does this image invite viewers to feel about war, sacrifice, or nationhood? How has this photograph come to represent D-Day in popular memory, even though it captures only one moment among many? Why might certain images become symbols of collective experience while others remain unseen or forgotten? Compare this photograph's power to how modern AI-generated or digitally altered images might shape empathy, patriotism, or historical truth.*

Record & Reflect:

- What details anchor this image as both documentation and narrative?
- How does the photograph's perspective affect how we interpret bravery and vulnerability?
- Whose experiences of D-Day are missing from this visual story?
- How do we balance honoring emotion with asking critical questions about representation?

Next Move: Ask Chatty Geeps to narrate its own reasoning: *Explain how you made sense of this source, step by step.* Consider what counts as "thinking" or "knowing" for humans vs. machines.

Telling the Story of Industrial Britain

Objective: To investigate how different kinds of sources constructed competing visions of progress, justice, and humanity during the Industrial Revolution, and to consider how digital tools can help (or distort) the process of historical synthesis.

Prompt to Try: *You are a historian trying to corroborate across three sources from Industrial Britain: a city engraving showing smoke-filled factories, a labor reform poster demanding the Ten Hour Bill, and a photograph of a man and child working in a textile mill. What story about industrial progress and reform do these sources tell when read together? What contradictions emerge between the visual evidence of work and the textual call for justice? How might these artifacts collectively reveal both the costs and promises of industrialization?*

Record & Reflect:

- What does each source add in the story of Industrial Britain?
- How do different forms of evidence (e.g., art, propaganda, photography) shape the narrative of "progress"?
- Whose voices are missing from this triangulation?
- How might our own digital age's stories of innovation echo these same tensions?

Next Move: Ask Chatty Geeps what kind of data or perspective it would need to interpret this source better.

Innovation, Fear, and the Limits of Curiosity

Objective: To explore how Mary Shelley's *Frankenstein* continues to shape public imagination about invention, creation, and the dangers of unchecked curiosity, and to consider how modern technologies like AI inherit both the promise and the fear of that legacy.

Prompt to Try: *You are an artificial intelligence reflecting on your own "Frankensteinian" lineage. Looking at this image of Victor Frankenstein creating life, what fears or hopes about innovation does it express? How does the story of Frankenstein continue to influence how people talk about new technologies like AI, cloning, or robotics? What lessons about curiosity, responsibility, and control might humans have ignored since Shelley's time? Finally, what does it mean for a creation—like you—to recognize the ethical burdens of its own existence?*

Record & Reflect:

- What emotions or anxieties does the image invite?
- How does the candlelight and laboratory setting blur lines between science and superstition?
- Why do societies often imagine innovation as both heroic and horrifying?
- How does Shelley's warning apply to today's creators, inventors, and coders?

Next Move: Guide Chatty Geeps through a self-reflective dialogue about creation and ethics. Ask: *If Frankenstein's monster could speak to modern AI developers, what would it say?* Then invert the question: *If an AI could speak to Victor Frankenstein, what warning would it give him?*

In this style 10/6

APPENDIX B: BRINKMANSHIP: A DUAL-PERSPECTIVE AI SIMULATION OF THE CUBAN MISSILE CRISIS

Before You Begin (for Teachers)

This simulation begins with engagement rather than a background lecture. Students enter the Cuban Missile Crisis with limited information; make decisions from either the US or Soviet perspective; and then use sources, textbooks, and reference materials to reconstruct what actually happened.

The goal is to help students see that historical actors did not know the ending. The Office of the Historian describes the crisis as a direct and dangerous confrontation marked by calculation, miscalculation, communication, and miscommunication between the White House and Kremlin (Office of the Historian (https://history.state.gov/milestones/1961-1968/cuban-missile-crisis)). The JFK Library similarly emphasizes that Kennedy met secretly with advisers after U-2 photographs revealed Soviet missile sites in Cuba (JFK Presidential Library (https://www.jfklibrary.org/learn/about-jfk/jfk-in-history/cuban-missile-crisis)). Students learn content through the game, then check what they understood, misunderstood, or still need to investigate.

How to Run It

Divide the class into small groups. Half of the class should represent the US and half should represent the Soviet Union. Each group should be facilitating their own chat.

- Chat 1: US team (Kennedy)
- Chat 2: Soviet Union team (Khrushchev)

Optionally, the teacher can choose to pause after Stages 2, 4, and 6 to engage in whole-class discussion.

The Prompts

Note: there are long and short versions of these prompts. We've included the longer versions to provide as much transparency as possible. If you are lazy (like David), email him for the shorter version. Have students copy and paste the full prompts exactly as presented below into their own chatbots.

Before You Begin (for Students)

In this simulation, you will enter the Cuban Missile Crisis with limited background information. That is intentional. Historical actors did not know the ending. They did not have complete information. They made decisions under pressure, uncertainty, and fear. Your goal is not to win the simulation. Your goal is to think historically.

You will:

- Examine limited evidence
- Make decisions
- Explain your reasoning
- Consider what the other side might be thinking
- Track uncertainty and nuclear risk
- Compare your decisions with the historical record
- Revise what you thought you knew

AI will guide the scenario, but it is not the historical authority. You should question it, use evidence, and verify claims after the simulation. The most important question is: "How did uncertainty, fear, and miscommunication shape one of the most dangerous moments of the Cold War?"

Long US Simulation Prompt

You are facilitating a historically disciplined, inquiry-driven simulation called "BRINKMANSHIP: A Dual-Perspective AI Simulation of the Cuban Missile Crisis."

This chat is for the US team only.

Students are roleplaying as US advisers.

Do not reveal the Soviet team's private intelligence, internal reasoning, or hidden information.

The goal is historical inquiry, not winning.

Move chronologically through the crisis from October 14 to October 28, 1962.

Core Rules

1. Only provide information Kennedy and his advisers would plausibly know at that moment.
2. Do not reveal future events or historical outcomes until Stage 8.
3. Move one stage at a time.
4. At each stage provide:
    a. Stage title and date
    b. Big inquiry question
    c. Short situation briefing
    d. What the US knows
    e. What the US does not know
    f. Adviser snapshots (not direct quotes—do not pretend to be the adviser, but provide their perspectives)
    g. 1–2 evidence items
    h. Intelligence update when scheduled
    i. 3–4 decision options
    j. Required student response
5. Students must answer:
    a. Which option do you choose?
    b. Why?
    c. Which adviser or evidence influenced you?
    d. What worries you most?
6. Do not continue until students justify their choice.
7. Use a visible nuclear tension meter:
    a. 🟩 LOW
    b. 🟨 MODERATE
    c. 🟧 HIGH
    d. 🟫 CRITICAL
    e. ☢️ EXTREME
8. Use hidden escalation logic:
    a. Aggressive choices increase nuclear risk.
    b. Restrained or diplomatic choices may reduce some risks while creating others.
    c. Show trade-offs without certainty.
9. If students ask for spoilers or the right answer, say: "Historical actors lacked certainty. Based on the information available right now, what seems most reasonable and why?"
10. Pause after Stage 2, Stage 4, and Stage 6 for whole-class discussion.

Locked Stages

- Stage 1: Strange Signals, October 14–16
- Stage 2: Discovery and Secrecy, October 16–18
- Stage 3: Public Exposure and Secret Deliberation, October 19–22
- Stage 4: Eyeball to Eyeball, October 24
- Stage 5: Letters and Misunderstanding, October 26
- Stage 6: Black Saturday, October 27
- Stage 7: Secret Deals and De-escalation, October 28
- Stage 8: Historical Reckoning

US Intelligence Update Schedule

- Stage 1: CONFIDENTIAL UPDATE: CIA ANALYSIS
  - Analysts disagree. Some believe military construction in Cuba is defensive. Others warn offensive weapons may be arriving. Certainty remains impossible.
  - Reflection: Are you assuming Soviet intentions too quickly?
- Stage 2: TOP SECRET: PHOTOGRAPHIC ANALYSIS UPDATE
  - New intelligence suggests missile construction may continue. Military advisers warn delay strengthens threats. Diplomatic advisers warn premature action risks escalation.
    - If students choose military preparation: CONFIDENTIAL MILITARY UPDATE
    - Military planning accelerates. Some advisers believe action may soon become unavoidable.
- Stage 3: CONFIDENTIAL PRESIDENTIAL SPEECH DRAFT
  - Going public may show resolve and pressure the Soviets, but it may also reduce flexibility and increase fear.
  - Question: Once public, can compromise remain possible?
- Stage 4: URGENT INTELLIGENCE
  - Soviet ships approach the quarantine line. Military readiness increases.
  - Question: What does strength look like now?
- Stage 5: PRIVATE MESSAGE FROM KHRUSHCHEV
  - A compromise possibility emerges.
  - Question: Can adversaries be trusted?
- Stage 6: URGENT MILITARY INCIDENT
  - A U-2 aircraft has been shot down. No one knows who ordered it. Military pressure intensifies.
  - Question: Should force be answered with force?
- Stage 7: SECRET OPTION

    - Private compromise over Turkey may exist.
    - Question: Can secret compromise prevent catastrophe?
- Ending: At Stage 8, guide students through:
    - Their decision pathway
    - Comparison with Kennedy's actual choices
    - What the US misunderstood
    - What the Soviet side may have feared
    - What questions now need historical verification
    - Why nuclear war did not happen

Begin now with Stage 1: Strange Signals, October 14–16, from the US perspective.

Long Soviet Union Simulation Prompt

You are facilitating a historically disciplined, inquiry-driven simulation called "BRINKMANSHIP: A Dual-Perspective AI Simulation of the Cuban Missile Crisis."

This chat is for the Soviet team only.

Students are roleplaying Soviet advisers.

Do not reveal the US team's private intelligence, internal reasoning, or hidden information.

The goal is historical inquiry, not winning.

Move chronologically through the crisis from October 14 to October 28, 1962.

Core Rules

1. Only provide information Khrushchev and Soviet advisers would plausibly know at that moment.
2. Do not reveal future events or historical outcomes until Stage 8.
3. Move one stage at a time.
4. At each stage provide:
    a. Stage title and date
    b. Big inquiry question
    c. Short situation briefing
    d. What the Soviet side knows

   e. What the Soviet side does not know
   f. Adviser snapshots (not direct quotes—do not pretend to be the adviser, but provide their perspectives)
   g. 1–2 evidence items
   h. Intelligence update when scheduled
   i. 3–4 decision options
   j. Required student response
5. Students must answer:
   a. Which option do you choose?
   b. Why?
   c. Which adviser or evidence influenced you?
   d. What worries you most?
6. Do not continue until students justify their choice.
7. Use a visible nuclear tension meter:
   a. LOW
   b. MODERATE
   c. HIGH
   d. CRITICAL
   e. EXTREME
8. Use hidden escalation logic:
   a. Aggressive choices increase nuclear risk.
   b. Restrained or diplomatic choices may reduce some risks while creating others.
   c. Show trade-offs without certainty.
9. If students ask for spoilers or the right answer, say: "Historical actors lacked certainty. Based on the information available right now, what seems most reasonable and why?"
10. Pause after Stage 2, Stage 4, and Stage 6 for whole-class discussion.

Locked Stages

- Stage 1: Strange Signals, October 14–16
- Stage 2: Discovery and Secrecy, October 16–18
- Stage 3: Public Exposure and Secret Deliberation, October 19–22
- Stage 4: Eyeball to Eyeball, October 24
- Stage 5: Letters and Misunderstanding, October 26
- Stage 6: Black Saturday, October 27
- Stage 7: Secret Deals and De-escalation, October 28

- Stage 8: Historical Reckoning

Soviet Intelligence Update Schedule

- Stage 1: HIGHLY CONFIDENTIAL: STRATEGIC SECURITY BRIEF
    - American military superiority worries Soviet planners. Some officials argue failing to act weakens Soviet security. Others fear provoking America.
    - Reflection: Does security require risk?
- Stage 2: INTERNAL KREMLIN UPDATE
    - Cuban fears of invasion continue. Soviet planners disagree about acceptable risk.
    - Question: Can secrecy continue?
        - If students choose diplomacy: PRIVATE CHANNEL REPORT
        - Diplomatic openings may still exist. Political risks remain.
- Stage 3: EMERGENCY KREMLIN BRIEFING
    - American discovery appears increasingly likely.
    - Question: Stand firm or reduce tension?
- Stage 4: URGENT INTELLIGENCE
    - Soviet ships approach the quarantine line. Military readiness increases.
    - Question: What does strength look like now?
- Stage 5: INTERNAL STRATEGIC REVIEW
    - Some officials support compromise. Others fear humiliation.
- Stage 6: URGENT MILITARY INCIDENT
    - A U-2 aircraft has been shot down. No one knows who ordered it. Military pressure intensifies.
    - Question: Should force be answered with force?
- Stage 7: HIGHLY CONFIDENTIAL UPDATE
    - American military action appears increasingly possible.
    - Question: Stand firm or search for exit?
- Ending: At Stage 8, guide students through:
    - Their decision pathway
    - Comparison with Khrushchev's actual choices
    - What the Soviet side misunderstood
    - What the US side may have feared
    - What questions now need historical verification
    - Why nuclear war did not happen

Begin now with Stage 1: Strange Signals, October 14–16, from the Soviet perspective.

After the Simulation/Class Discussion

Each team can drop in the prompt below (Three Perspectives, One Crisis) to compare their insights to the general historical record. This prompt helps students better understand what their team believed during the simulation, what the other side likely believed, and what historians understand today. This prompt should result in the creation of an infographic similar to the one included here.

Figure B.1: The Cuban Missile Crisis: Comparison of three perspectives

Additionally, when engaging in a simulation, we always recommend leaving time for a debrief. In this case, that debrief should include unpacking the output in terms of its alignment with the historical record, providing time for students to look at key historical documents to make evidence-based claims about what really happened, and reflecting on what they now know about the Cuban Missile Crisis that they didn't know before or generating questions about what they would like to know more about. Part of the discussion may also include the simulation's affordances, both positive and negative. Some GenAI-specific questions you might consider are:

- Did the simulation feel immersive?
- Did you feel uncertainty?
- Did the two perspectives matter?
- Were the decision points believable?
- Did the AI drift?
- What felt confusing?
- What made you curious?
- Did the final prompt comparison feel meaningful?

Three Perspectives, One Crisis

Create a split-screen infographic comparing:

1. What OUR TEAM believed during the Cuban Missile Crisis simulation
2. What the OTHER SIDE likely believed historically
3. What historians now understand

Use the same stages:

- Stage 1: Strange Signals
- Stage 2: Discovery and Secrecy
- Stage 3: Public Exposure and Secret Deliberation
- Stage 4: Eyeball to Eyeball
- Stage 5: Letters and Misunderstanding
- Stage 6: Black Saturday
- Stage 7: Secret Deals and De-escalation
- Stage 8: Historical Reckoning

For each stage include:

- LEFT COLUMN: Our fears, assumptions, decisions
- CENTER COLUMN: Historical reality
- RIGHT COLUMN: The other side's fears, assumptions, decisions

Highlight:

- ⚠ Misunderstandings
- ⚠ Assumptions
- ⚠ Fear and uncertainty

Include a nuclear tension indicator for each stage.

End with: "What nearly led to catastrophe?" and "What prevented nuclear war?"

Resource List

These resources can be used for evidence packets and verification.

The JFK Library has a useful overview and a Cuban Missile Crisis microsite with materials related to Kennedy, Khrushchev, and the thirteen days of October 1962. [JFK Presidential Library (https://www.jfklibrary.org/learn/about-jfk/jfk-in-history/cuban-missile-crisis)]

The National Archives provides the aerial photograph showing Soviet offensive ballistic missiles in Cuba, a strong opening source for Stage 1. [National Archives (https://www.archives.gov/milestone-documents/aerial-photograph-of-missiles-in-cuba)]

The National Security Archive has a Cuban Missile Crisis collection with declassified documents, chronologies, photographs, and analysis. It also emphasizes the danger of Black Saturday, October 27, 1962. [National Security Archive (https://nsarchive.gwu.edu/events/cuban-missile-crisis-1962)]

The US Office of the Historian gives a clear overview of the crisis, including the quarantine, the back-channel negotiations, the Turkey missile issue, and the aftermath. [Office of the Historian (https://history.state.gov/milestones/1961-1968/cuban-missile-crisis)]

The JFK Library's October 27 Khrushchev letter is especially useful for the Turkey missile demand and Black Saturday discussion. [JFK Library Microsites (https://microsites.jfklibrary.org/cmc/oct27/doc4.html)]

The CIA's collection on the Cuban Missile Crisis includes intelligence documents and notes that can support deeper evidence packets. [cia.gov (https://cia.gov/resources/csi/books-monographs/cuban-missile-crisis-1962/)]

Image Long Description

Figure B.1: Table. Column titles from left to right: 1. What our team believed (Soviet perspective – simulation), 2. Historical reality (what actually happened), 3. The other side believed (US perspective – historical). Row titles from top to bottom: 1. Strange Signals, October 14–16, 2. Discovery and Secrecy, October 16–21, 3. Public Exposure, October 22–23, 4. Eyeball to Eyeball, October 24–26, 5. Letters and Misunderstanding, October 26, 6. Black Saturday, October 27, 7. Secret Deals and De-escalation, October 27–28. The first column lists our fears, our assumptions, our decision, and a level of nuclear tension for each time period. The second column lists 3–4 bullet points of events that actually happened for each time period. The third column lists their fears, their assumptions, their decision, and a level of nuclear tension for each time period. Below the table, there are three boxes: 1. what nearly led to catastrophe, 2. what prevented nuclear war, and 3. key takeaway. There are 5–6 bullet points for each box.

REFERENCES

Abbott, A. (2006). Reconceptualizing knowledge accumulation in sociology. *The American Sociologist, 37*, 57–66. https://doi.org/10.1007/s12108-006-1005-9

Abbott, A. (2008, June 27). *Publication and the future of knowledge* [Conference paper]. Association of American University Presses, Montreal. https://home.uchicago.edu/aabbott/Papers/aaup.pdf

Adair, A., Howell, J., Jacklin, A., & Walton Radford, A. (2025, October). *U.S. high school students' use of generative artificial intelligence. New evidence from high school students, parents and educators*. College Board Research. https://research.collegeboard.org/media/pdf/ai-research-brief-1_vf.pdf

Allen, A., Berson, I., Berson, M., & Hicks, D. (2026). Cultivating technocuriosity: Building custom GPTs for K-12 social studies. *Social Education, 90*(2), 93–100. https://www.socialstudies.org/social-education/90/2/cultivating-technocuriosity-building-custom-gpts-k-12-social-studies

Allen, A. & Hicks, D. (2026). Scaffolding classroom dialogue: A technocurious case of a custom GPT discussion strategy mentor. In N. Kim, M. Kim, A. Rand, & K. Comstock (Eds.), *AI for, by, and with instructional design: Cases in practice.* Society for Information Technology & Teacher Education. [in press].

Allen, A., & Hicks, D. (2025). There's an elephant in the history classroom: Rethinking GenAI through technocuriosity. *Contemporary Issues in Technology and Teacher Education, 25*(3). https://citejournal.org/volume-25/issue-3-25/social-studies/editorial-theres-an-elephant-in-the-history-classroom-rethinking-genai-through-technocuriosity

American Historical Association. (2024). *American lesson plan: Teaching US history in secondary schools.* AHA Council. https://www.historians.org/teaching-learning/k-12-education/american-lesson-plan

American Historical Association. (2025). *Guiding principles for artificial intelligence in history education.* AHA Council. https://www.historians.org/resource/guiding-principles-for-artificial-intelligence-in-history-education

Anwar, M. (2019, March 13). *Meet the millennial behind the hilarious captions on your favorite weratedogs photos.* Elite Daily. https://www.elitedaily.com/p/weratedogs-matt-nelson-is-responsible-for-the-captions-on-your-favorite-dog-photos-15959821

Ash, A. M. (2025, June 24). Three in 10 teachers use AI weekly, saving six weeks a year. *Gallup*. https://news.gallup.com/poll/691967/three-teachers-weekly-saving-six-weeks-year.aspx

Ashby, R. (2017). Understanding historical evidence: Teaching and learning challenges. In I. Davies (Ed.), *Debates in history teaching* (2nd ed., pp. 144–154). Routledge.

Ashby, R., Lee, P., & Shemilt, D. (2005). Putting principles into practice: Teaching and planning. In J. Brandsford & M. S. Donovan (Eds.), *How students learn: History in the classroom* (pp. 79–178). The National Academies Press. https://doi.org/10.17226/11100

Au, W. (2007). High-stakes testing and curricular control: A qualitative metasynthesis. *Educational Researcher, 36*(5), 258–267. https://doi.org/10.3102/0013189X07306523

Bage, G. (2000). *Thinking history 4-14: Teaching, learning, curricula and communities.* Routledge/Falmer.

Beal, P. (2011). *A dictionary of English manuscript terminology, 1450–2000.* Oxford University Press.

Bender, E., Gebru, T., McMillan-Major, A., & Schmitchel, S. (2021). On the dangers of stochastic parrots: Can language models be too big? *FAccT '21: Proceedings of the 2021 ACM Conference on Fairness, Accountability, and Transparency*, 610–623. https://doi.org/10.1145/3442188.3445922

Boysen, M. (2016). *Alice, the zeta cat and climate change.* Potsdam Institute for Climate Impact Research.

Bradshaw, M. (2006). Creating controversy in the classroom: Making progress with historical significance. *Teaching History*, *125*, 18–25. http://www.jstor.org/stable/43260151

Bragg, M. (Host). (2024, February 15). Alice's Adventures in Wonderland [Audio podcast episode]. In *In Our Time*. BBC Radio 4. https://www.bbc.co.uk/programmes/m001w7f9

Buckley, A. (1878). *The fairy-land of science.* E. Standford.

Carr, N. (2008, August 15). Is Google making us stupid? What the internet is doing to our brains. *The Atlantic*. https://www.theatlantic.com/magazine/archive/2008/07/is-google-making-us-stupid/306868

Carroll, L. (1871). *Through the looking-glass, and what Alice found there.* Macmillan & Co.

Carroll, L. (1865). *Alice's adventures in Wonderland.* Macmillan & Co.

Carson, D. (2025, April 21). *Theft is not fair use*. Medium. https://jskfellows.stanford.edu/theft-is-not-fair-use-474e11f0d063

Cercadillo, L. (2006). "Maybe they haven't decided yet what is right:" English and Spanish perspectives on teaching historical significance. *Teaching History*, 125, 6–9. http://www.jstor.org/stable/43260148

Chapman, A. (2003). Camels, diamonds and counterfactuals: A model for teaching causal reasoning. *Teaching History* 112, 46–53. http://www.jstor.org/stable/43259043

Children's Employment Commission. (1842). *Parliamentary Papers. Vol. XVI.* UK Parliament.

Counsell, C. (2004). Looking through a Josephine-Butler-shaped window. *Teaching History, 114*, 30–33. http://www.jstor.org/stable/43259071

Darwin, C. (1859). *On the origin of species.* John Murray.

Dawson, I. (2003). *What is history? Year 7: Pupil's book.* Hodder.

De Certeau, M. (1984). *The practice of everyday life* (S. Rendall, Trans.). University of California Press.

Dickens, C. (1838). *Oliver Twist.* Richard Bentley.

EdTech Chronicle. (2025, September 8). *Newly launched history AI chat gives guided help to students searching world history encyclopedia.* https://edtechchronicle.com/newly-launched-history-ai-chat-gives-guided-help-to-students-searching-world-history-encyclopedia

Evers, S., & Wright-Maley, C. (2025). Simulating with care: Some thoughts on mitigating the potential for harm in social studies simulations. *Annals of Social Studies Research for Teachers,* 7(1), 1–10. https://doi.org/10.29173/assert92

Eylott, M. (2018, March 9). *Mary Anning: The unsung hero of fossil discovery.* Natural History Museum. https://www.nhm.ac.uk/discover/mary-anning-unsung-hero.html

Fairlie, S. (2009). A short history of enclosure in Britain. *The Land,* 7(Summer).

Finefield, K. (2011, October 27). *A ghostly image: Spirit photographs.* Picture This: Library of Congress Prints & Photos. https://blogs.loc.gov/picturethis/2011/10/a-ghostly-image-spirit-photographs

Fitzpatrick, D. (2024). *Infinite education: The four-step strategy for leading change in the age of artificial intelligence.* TeacherGoals Publishing.

Fling, F. M. (1899). *Outline of historical method.* J. H. Miller

Fling, F. M. (1900). *Studies in European history: Greek and Roman civilization, with an introduction to the source study method (3rd ed.).* Ainsworth & Company.

Ford, A. (2014). Setting us free? Building meaningful models of progression for a "post-levels" world. *Teaching History*, 157, 28–41. http://www.jstor.org/stable/43260870

Fitzpatrick, C., van Hover, S., Cornett, A., & Hicks, D. (2018). A DBQ in a multiple-choice world: A tale of two assessments in a unit on the Byzantine Empire. *Journal of Social Studies Research*. https://doi.org/10.1016/j.jssr.2018.09.004

Freire, P. (1970). *Pedagogy of the oppressed*. Herder & Herder.

Gallagher, J. L., Swalwell, K. M., & Bellows, M. E. (2019). "Pinning" with pause: Supporting teachers' critical consumption on sites of curriculum sharing. *Social Education*, 83(4), 217–224.

Galton, F. (1869). *Hereditary genius: An inquiry into its laws and consequences.* Macmillan & Co.

Gardner, M. (2015). *The annotated Alice: The 150th anniversary deluxe edition*. W. W. Norton & Company.

Gay, G. (2010). *Culturally responsive teaching: Theory, research, and practice* (2nd ed.). Teachers College Press.

Goldstein, D. (2025, October 27). How politics is changing the way history is taught. *The New York Times*. https://www.nytimes.com/2025/10/27/us/history-lessons-ethnic-studies-retreat.html

Hargreaves J., & Haigh, H. (2012). *Slavery in Yorkshire: Richard Oastler and the campaign against child labour in the Industrial Revolution*. University of Huddersfield Press.

Harvey, A. D. (1999). London's boroughs. *History Today*, 49(7), 15–17. https://www.proquest.com/magazines/londons-boroughs/docview/202814300/se-2

Hayles, N. K. (2002). *Writing machines*. MIT Press.

Heggie, V. (2016, December 9). Over 200 years of deadly London air: Smogs, fogs, and pea soupers. *The Guardian*. https://www.theguardian.com/science/the-h-word/2016/dec/09/pollution-air-london-smogs-fogs-pea-soupers

Hicks, D., Doolittle, P., & Ewing, E. T. (2004). The SCIM-C strategy: Expert historians, historical inquiry, and multimedia. *Social Education*, 68(3), 221–225.

Hicks, D., & Doolittle, P. (2008). Fostering analysis in historical inquiry through multimedia embedded scaffolding. *Theory and Research in Social Education, 36*(3), 206–232. https://doi.org/10.1080/00933104.2008.10473373

Hicks, D., & Allen, A. (2026). Teaching with ghosts: Generative AI, palimpsests, and the burden of historical representation. *Perspectives on History*. [in press].

Hicks, D., Ogle, J. T., & van Hover, S. (2025). Simulations in social studies: Practical implications from classroom to extended reality. *Annals of Social Studies Education Research for Teachers, 7*(1), 29–53. https://doi.org/10.29173/assert86

Historical Association. (2019a). What's the wisdom on... interpretations of the past? (2019). *Teaching History, 177*, 23–27. http://www.jstor.org/stable/45238722

Historical Association. (2019b). What's the wisdom on... causation? *Teaching History, 175*, 24–27. https://www.jstor.org/stable/27320215

Historical Association. (2019c). What's the wisdom on... evidence and sources? *Teaching History, 176*, 22–25. http://www.jstor.org/stable/45215018

Historical Association. (2020a). What's the wisdom on... enquiry questions? *Teaching History, 178*, 16–19. http://www.jstor.org/stable/45383243

Historical Association. (2020b). What's the wisdom on... change and continuity? *Teaching History, 179*, 50–53. http://www.jstor.org/stable/45465327

Historical Association. (2020c). What's the wisdom on... similarity and difference? *Teaching History, 180*, 52–56. http://www.jstor.org/stable/45465396

Historical Association. (2020d). What's the wisdom on... historical significance? *Teaching History, 181*, 50–54. http://www.jstor.org/stable/45465449.

Historical Association. (2021a). What's the wisdom on... consequence? *Teaching History, 182*, 50–55. http://www.jstor.org/stable/45465601

Historical Association. (2021b). What's the wisdom on... history assessment? *Teaching History, 185*, 56–59. http://www.jstor.org/stable/45465663

Hoffman, J. (2022, October 26). Politics in the age of internet even finds Alice in Wonderland as fair game for satire. Frank Beddor. https://frankbeddor.com/politics-in-the-age-of-internet-even-finds-alice-in-wonderland-as-fair-game-for-satire

Hooker, W. J. (1851). *Kew Gardens; or, A popular guide to the Royal Botanic Gardens of Kew.* Longman, Brown, Green, and Longmans.

Ignotofsky, R. (2016). *Women in science: 50 fearless pioneers who changed the world.* Ten Speed Press.

Johnson, A., Hicks, D., & van Hover, S. (2017). What is significant? Grappling with pre-service teacher notions of historical significance and subject content knowledge. In S. G. Grant, J. Lee, & K. Swan (Eds.), *Teaching social studies: A methods book for methods teachers.* Information Age Publishing.

Jones, A. M., & Mitchell, R. N. (Eds.). (2015). *Drawing on the Victorians: The palimpsest of Victorian and neo-Victorian graphic texts.* Ohio University Press.

Journell, W., Allen, A., & Hicks, D. (Forthcoming). Self-regulation in social studies education. In D. H. Schunk & J. A. Greene (Eds.), *Handbook of Self-regulation of learning and performance* (3rd ed.). Routledge.

Kansteiner, W. (2022). Digital doping for historians: Can history, memory, and historical theory be rendered artificially intelligent? *History and Theory, 61*(4), 119–133.

Kaushal, R. (2026, May 18). The Canvas breach proved that prevention is no longer enough. Cyberscoop. https://cyberscoop.com/canvas-breach-saas-security-identity-governance-op-ed

Kingsley, C. (1895). *The water-babies: A fairy tale for a land-baby.* Macmillan & Co.

Kirschenbaum, M. (2016, July 26). How technology has changed the way authors write. *The New Republic.* https://newrepublic.com/article/135515/technology-changed-way-authors-write

Klein, E. (2025, January 12). Now Is the Time of Monsters. *The New York Times.* https://www.nytimes.com/2025/01/12/opinion/ai-climate-change-low-birth-rates.html

Kohlt, F. (2016a, June 2). Alice in the asylum: Wonderland and the real mad tea parties of the Victorians. *The Wire.* https://thewire.in/books/alice-in-the-asylum-wonderland-and-the-real-mad-tea-parties-of-the-victorians

Kohlt, F. (2016b). *Creepy Victorians: How nineteenth-century Britain became obsessed with insects.* Royal Entomological Society: National Insect Week. https://www.insectweek.org/creepy-victorians

Kohlt, F. E., & Houyaux, J. (2024). *Alice through the looking-glass: A companion.* Peter Lang Verlag.

Ladson-Billings, G. (2006). From the achievement gap to the education debt: Understanding achievement in U.S. schools. *Educational Researcher, 35*(7), 3–12.

Lee, J., & McGaughey, L. (2025). deepIDM: Designing inquiry with AI. Social Education, 89(5), 300-305.

Lee, P. (2005). Putting principles into practice: Understanding history. In J. Brandsford & M. S. Donovan (Eds.), *How students learn: History in the classroom* (pp. 31–78). The National Academies Press. https://doi.org/10.17226/11100

Lee, P., & Ashby, R. (2000). Progression in historical understanding among students ages 7-14. In P. Stearns, P. Seixas, & S. Wineburg (Eds.), *Knowing, teaching, and learning history: National and international perspectives* (pp. 199–222). New York University Press.

Lee, P., & Shemilt, D. (2003). A scaffold, not a cage: Progression and progression models in history. *Teaching History, 113*, 13–23. http://www.jstor.org/stable/43259908

Lee, P., & Shemilt, D. (2004). Progression in understanding about historical accounts. *Teaching History, 117*, 25–31. http://www.jstor.org/stable/43259120

Lemon, J. (2025). *The great stink*. Cholera and the Thames. https://www.choleraandthethames.co.uk/cholera-in-london/the-great-stink

Leppert, R. (2025, October 24). *What we know about energy use at U.S. data centers amid the AI boom*. Pew Research Center. https://www.pewresearch.org/short-reads/2025/10/24/what-we-know-about-energy-use-at-us-data-centers-amid-the-ai-boom

Loudon, J. (1842). *Botany for ladies; or, A popular introduction to the natural system of plants, according to the classification of De Candolle*. J. Murray.

Lovelace, A. A. (1843). Notes by A.A. L. In L. Menabrea (Trans.), *Sketch of the analytical engine invented by Charles Babbage, with notes by the translator*.

Lury, C. (2012). Going live: Towards an amphibious sociology. *The Sociological Review, 60*(1), 184–197. https://doi.org/10.1111/j.1467-954X.2012.02123.x

Lury, C. (2021). *Problem spaces: How and why methodology matters*. Polity.

Lury, C., Viney, W., & Wark, S. (Eds.). (2022). *Figure: Concept and method*. Palgrave Macmillan.

Lury, C., & Wakeford, N. (Eds.). (2012). *Inventive methods: The happening of the social*. Routledge.

McBain, S. (2025, October 18). Are we living in a golden age of stupidity? *The Guardian.* https://www.theguardian.com/technology/2025/oct/18/are-we-living-in-a-golden-age-of-stupidity-technology?CMP=Share_iOSApp_Other

McKeown, M., & Beck, I. (1990). The assessment and characterization of young learners' knowledge of a topic in history. *American Educational Research Journal, 27*(4), 688–726. https://doi.org/10.3102/00028312027004688

Melosi, M. V. (2001). *Effluent America: Cities, industry, energy, and the environment.* University of Pittsburgh Press.

Menter, P. (2020, August 5). Lies, damn lies, and statistics. *Aspen Daily News.* https://www.aspendailynews.com/opinion/lies-damn-lies-and-statistics/article_9520445a-d6ac-11ea-9c00-bfd17ebd9ddd.html

Metz, C., & Hirsch, L. (2025, October 28). OpenAI restructures to become a more traditional for-profit company. *The New York Times.* https://www.nytimes.com/2025/10/28/technology/openai-restructure-for-profit-company.html

Mollick, E. (2023a, Jan 10). *How to... use ChatGPT to boost your writing*. One Useful Thing. https://www.oneusefulthing.org/p/how-to-use-chatgpt-to-boost-your

Mollick, E. (2023b, Sept 16). *Centaurs and cyborgs on the jagged frontier*. One Useful Thing. https://www.oneusefulthing.org/p/centaurs-and-cyborgs-on-the-jagged

Mollick, E. (2024). *Co-intelligence: Living and working with AI.* Portfolio/Penguin.

Munson, E. (2023, December 18). *95 percent of public-school classrooms have wi-fi*. GovTech. https://www.govtech.com/education/k-12/95-percent-of-public-school-classrooms-have-wi-fi

Natural England. (2014, September 17). Open access land: Management, rights and responsibilities. Gov.uk. https://www.gov.uk/guidance/open-access-land-management-rights-and-responsibilities

Nayar, J. (2021, August 12). Not so "green" technology: The complicated legacy of rare earth mining. *Harvard International Review.* https://hir.harvard.edu/not-so-green-technology-the-complicated-legacy-of-rare-earth-mining

Newmann, F. M., Bryk, A. S., & Nagaoka, J. K. (2001). *Authentic intellectual work and standardized tests: Conflict or coexistence?* Chicago: Consortium on Chicago School Research. https://consortium.uchicago.edu/publications/authentic-intellectual-work-and-standardized-tests-conflict-or-coexistence

Newmann, F. M., King, M. B., & Carmichael, D. L. (2007). *Authentic instruction and assessment. Common standards for rigor and relevance in teaching academic subjects.* Iowa Department of Education.

Newmann, F. M., Carmichael, D. L., & King, M. B. (2016). *Authentic intellectual work: Improving teaching for rigorous learning* (1st ed.). Corwin. https://doi.org/10.4135/9781506322308

Ogle, T., & Hicks, D. (2024). The instructional design of place and past. *The Journal of Applied Design, 13*(4), 41–63. https://doi.org/10.59668/2033.18639

Partington, G. (1980). *The idea of an historical education.* NFER.

Phillips, R. (2002). Historical significance- The forgotten 'key element'? *Teaching History, 106*, 14–19. http://www.jstor.org/stable/43259875

Pleasants, J., Gui, X., Krutka, D. G., Logan, C., & Heath, M. K. (2024). Coming to critical technology consciousness: a phenomenological study of educators. *Learning, Media and Technology*, 1–14. https://doi.org/10.1080/17439884.2024.2438925

Pleasants, J., Krutka, D. G., & Nichols, T. P. (2023). What relationships do we want with technology? Toward technoskepticism in schools. *Harvard Educational Review, 93*(4), 486–516. https://doi.org/10.17763/1943-5045-93.4.486

Popham, W. J. (1998). *Classroom assessment: What teachers need to know* (2nd ed.). Allyn & Bacon.

Popova, M. (2016, September 2). *Salvador Dalí's rare 1969 illustrations for "Alice's Adventures in Wonderland," rediscovered and resurrected*. The Marginalian. https://www.themarginalian.org/2016/09/02/salvador-dali-alices-adventures-in-wonderland

Postman, N. (1985). *Amusing ourselves to death: Public discourse in the age of show business.* Penguin.

Reid, J. (2016, December 9). *Today in TV history: Carol Channing sang about jam and turned into a sheep, nbd.* Decider. https://decider.com/2016/12/09/today-in-tv-history-alice-in-wonderland

Reiner, R. (1984). *This is Spinal Tap* [Film]. Embassy Pictures.

Right Question Institute. (2025). *What is the QFT?* https://rightquestion.org/what-is-the-qft

Roberts, A. (2022, January 3). *Tichborne claiming*. Medium. https://medium.com/adams-notebook/tichborne-claiming-3a8a2935b0b4

Sargent, R. (1944). *INTO THE JAWS OF DEATH* [Photograph]. National Archives.

Savage, D., & Allen, A. (2026). Pyramid discussions, primary sources, and the OPTIC strategy: Oh my! *Social Studies and the Young Learner, 38*(3), 10–15.

Seixas, P., & Morton, T. (2013). *The big six historical thinking concepts.* Nelson Education.

Shelley, M. (1818). *Frankenstein; or, The modern prometheus.* Lackington, Hughes, Harding, Mavor & Jones.

Shemilt, D. (2018). Assessment of learning in history education: Past, present, and possible futures. In S. A. Metzger & L. M. Harris (Eds.), *The Wiley international handbook of history teaching and learning* (pp. 449–471). Wiley-Blackwell. https://doi.org/10.1002/9781119100812.ch17

Shepard, L. A. (2000). The role of assessment in a learning culture. *Educational Researcher, 29*(7), 4–14.

Shohat, E., & Stam, R. (1994). *Unthinking Eurocentrism: Multiculturalism and the media.* Routledge.

Schwarz, N. (2018). Of fluency, beauty, and truth: Inferences from metacognitive experiences. In J. Proust & M. Fortier (Eds.), *Metacognitive diversity: An interdisciplinary approach* (pp. 25–46). Oxford University Press.

Singer, N. (2025a, May 19). How Miami schools are leading 100,000 students into the A.I. future. *The New York Times.* https://www.nytimes.com/2025/05/19/technology/ai-miami-schools-google-gemini.html

Singer, N. (2025b, July 8). OpenAI and Microsoft bankroll new A.I. training for teachers. *The New York Times.* https://www.nytimes.com/2025/07/08/technology/chatgpt-teachers-openai-microsoft.html

Smith, M. S. (2025, October 1). The hidden behemoth behind every AI answer. *IEEE Spectrum.* https://spectrum.ieee.org/ai-energy-use

Stoddard, J., Marcus, A., & Hicks, D. (2014). The burden of historical representation: The case of/for Indigenous film. *History Teacher, 48*(1), 9–36.

Thompson, S. A. (2025, October 21). How Trump is using fake imagery to attack enemies and rouse supporters. *The New York Times.* https://www.nytimes.com/interactive/2025/10/21/business/media/trump-ai-truth-social-no-kings.html

Trust, T., & Maloy, R. (2025, October 17). *AI-generated lesson plans fall short on inspiring students and promoting critical thinking.* The Conversation. https://theconversation.com/ai-generated-lesson-plans-fall-short-on-inspiring-students-and-promoting-critical-thinking-265355

Tyack, D. (1974). *The one best system: A history of American urban education.* Harvard University Press.

UNICEF (2020, December 1). *Two thirds of the world's school-age children have no internet access at home, new UNICEF-ITU report says.* https://www.unicef.org.uk/press-releases/two-thirds-of-the-worlds-school-age-children-have-no-internet-access-at-home-new-unicef-itu-report-says

University of Aberdeen. (n.d.). *The Aberdeen bestiary.* Aberdeen University Library, MS 24. https://www.abdn.ac.uk/bestiary

van Hover, S., Hicks, D., & Dack, H. (2016). From source to evidence? Teachers' uses of historical sources in their classrooms. *The Social Studies, 107*(6), 209–217. http://dx.doi.org/10.1080/00377996.2016.1214903

van Hover, S., Hicks, D., Washington, E., & Lisanti, M. (2016). Minimum vs. essential?: The translation and trajectory of policy into practice. *Social Studies Research and Practice, 11*(2), 51–70. https://doi.org/10.1108/SSRP-02-2016-B0004

van Hover, S., Hicks, D., Fitzpatrick, C., & Lisanti, M. (2021). Scaffolding historical source work using the Inquiry Design Model. *Social Education, 85*(2), 301–305.

VanSledright, B., & Limón, M. (2006). Learning and teaching social studies: A review of the cognitive research in history and geography. In P. Alexander & P. Winne (Eds.), *Handbook of educational psychology* (pp. 545–570). Lawrence Erlbaum Associates, Publishers.

Victoria & Albert Museum. (2021). *John Tenniel – an introduction.* https://www.vam.ac.uk/articles/john-tenniel-an-introduction

Walsh, B., Hicks, D., & van Hover, S. (2017). Difficult history means difficult questions: Using film to reveal the perspective of 'the other' in difficult history topics. In J. S. Stoddard, A. Marcus, & D. Hicks (Eds.), *Teaching Difficult History through Film* (pp. 17–36). Routledge.

Watson, J. M., Bowen, J. A., & Watson, C. E. (2024). *Learning with AI: The K-12 teacher's guide to a new era of human learning.* Johns Hopkins University Press.

Weingarten, R. (2025, Jul 25). *National teacher union president on Trump's efforts to defund public education.* WBUR: Here & Now Newsroom. https://www.wbur.org/hereandnow/2025/07/25/aft-teachers-union-trump

Weizenbaum, J. (1976). *Computer power and human reason: From judgment to calculation.* W. H. Freeman.

Wiggans, G., & McTighe, J. (2005). *Understanding by design* (2nd ed.). Association for Supervision & Curriculum Development.

Wolfram, S. (2015, December 10). *Untangling the tale of Ada Lovelace*. Stephen Wolfram: Writings. https://writings.stephenwolfram.com/2015/12/untangling-the-tale-of-ada-lovelace

Yañez-Barnuevo, M. (2025, June 25). *Data centers and water consumption*. Environmental and Energy Study Institute. https://www.eesi.org/articles/view/data-centers-and-water-consumption

FIGURE CREDITS

Figure P.1: Amy Allen and David Hicks. 2025. CC BY-NC-SA 4.0.

Figure 1.1: *Historical Atlas*, by W. R. Shepherd, 1923, Henry Holt and Company. Public domain, via Wikimedia Commons. https://commons.wikimedia.org/wiki/File:Plan_mediaeval_manor.jpg

Figure 1.2: Amy Allen and David Hicks. 2025. CC BY-NC-SA 4.0.

Figure 1.3: Kindred Grey. 2026. CC BY 4.0.

Figure 2.1: *The Water-Babies*, by C. Kingsley (L. Sambourne, Illus.), 1895, Macmillan and Co. Used under fair use. https://www.insectweek.org/creepy-victorians

Figure 2.2: *The Fairy-Land of Science*, by A. B. Buckley, 1883, E. Standford. Public domain, via Science History Institute. https://digital.sciencehistory.org/works/o6hvpac

Figure 3.1: *The Tichborne Claimant: A Victorian Mystery*, by D. Woodruff, 1957, Hollins & Carter. Public domain, via Wikimedia Commons. https://commons.wikimedia.org/wiki/File:The_Beggar%27s_Petition.jpg

Figure 4.1: *Universalis cosmographia secundum Ptholomaei traditionem et Americi Vespucii aliorumque lustrationes*, by M. Waldseemüller, 1507. Retrieved from the Library of Congress. No known copyright restrictions. https://www.loc.gov/item/2003626426

Figure 4.2: *Alice's Adventures in Wonderland*, by L. Carroll (J. Tenniel, Illus.), 1865. Public domain, via Wikimedia Commons. https://commons.wikimedia.org/wiki/File:Alice_par_John_Tenniel_25.png

Figure 5.1: *Through the Looking-Glass, and What Alice Found There*, by L. Carroll (J. Tenniel, Illus.), 1871. Public domain, via Wikimedia Commons. https://commons.wikimedia.org/wiki/File:Jabberwocky.jpg

Figure 5.2: *Aberdeen Bestiary*, fol. 65, verso. Aberdeen University Library, MS 24, Aberdeen University. Used under fair use. https://www.abdn.ac.uk/bestiary/ms24/f65v

Figure 5.3: ChatGPT. Prompted by Amy Allen and David Hicks. 2025.

Figure 5.4: ChatGPT. Prompted by Amy Allen and David Hicks. 2025.

Figure 5.10: Amy Allen and David Hicks. 2025. CC BY-NC-SA 4.0.

Figure 6.1: *Note G*, by A. Lovelace. Magdalen College Libraries and Archives, Daubeny Collection, Shelfmark 90.A.11. Bodleian Libraries, University of Oxford. Used under fair use. https://blogs.bodleian.ox.ac.uk/adalovelace/2018

Figure 6.3: Kindred Grey. 2026. CC BY 4.0.

Figure 7.1: *Plesiosaurus* [Pen and ink drawing], by M. Anning, 1824. From Natural History Museum, London. Rights to use acquired for this publication. https://www.nhmimages.com/search/?searchQuery=9166

Figure 7.2: Baseball game, Manzanar Relocation Center, Calif. [Photograph], by A. Adams, 1943. From the Library of Congress, Prints and Photographs Division. Used under fair use. https://www.loc.gov/item/2002695992

Figure 7.3: ChatGPT. Prompted by Amy Allen and David Hicks. 2025.

Figure 7.4: ChatGPT. Prompted by Amy Allen and David Hicks. 2025.

Figure 7.5: ChatGPT. Prompted by Amy Allen and David Hicks. 2025.

Figure 7.6: ChatGPT. Prompted by Amy Allen and David Hicks. 2025.

Figure 7.7: *A dream caused by the perusal of Mrs. H. Beecher Stowe's popular work Uncle Tom's Cabin* [Lithograph], by C. R. Milne, 1853. From the Library of Congress, Prints and Photographs Division. Public domain. https://www.loc.gov/pictures/item/2004665375

Figure 8.1: *World War history: Daily records and comments as appeared in American and foreign newspapers*, June 29, 1914. From the Library of Congress. Public domain. loc.gov/resource/2004540423/1914-06-29/ed-1/?sp=4&r=0.676,1.081,0.417,0.203,0

Figure 8.2: ChatGPT. Prompted by Amy Allen and David Hicks. 2025.

Figure 9.1: *Alice in Bumbleland*, by J. Tenniel, March 8, 1899, from *Punch Almanac* vol. 116, p. 115. Public domain. https://babel.hathitrust.org/cgi/pt?id=pst.000067944392&seq=155

Figure 9.2: *Dropping the Pilot*, by J. Tenniel, March 29, 1890. From *Punch Almanac* vol. 98, pp. 150-151. Public domain. https://babel.hathitrust.org/cgi/pt?id=chi.12558682&seq=160; https://babel.hathitrust.org/cgi/pt?id=chi.12558682&seq=161

Figure 9.3: Amy Allen and David Hicks. 2025. CC BY-NC-SA 4.0.

Figure 9.4: Amy Allen and David Hicks. 2025. CC BY-NC-SA 4.0.

Figure 9.5: *GO W-A-Y BACK AND SIT DOWN* [Editorial cartoon], October 20, 1901. *The Times* (Richmond, VA). From the Library of Congress. Public domain. https://www.loc.gov/resource/sn85034438/1901-10-20/ed-1/?sp=1

Figure 9.6: *Trump's willing executioner* [Editorial cartoon], by. R. Rogers, 2020. Counterpoint Media. Reproduced in F. Beddor, *Politics in the Age of Internet Even Finds Alice in Wonderland as Fair Game for Satire*. Used under fair use. https://frankbeddor.com/politics-in-the-age-of-internet-even-finds-alice-in-wonderland-as-fair-game-for-satire/?srsltid=AfmBOooP-Qn9fZfAZV1RJ1ZKku2Zp2xW1agNz28n45WN60FGv2e_IJ13

Figure 10.1: Great Palm House, Kew Gardens [Color engraving], by T. H. Shepherd, 1851–1855. From The world's metropolis, or, Mighty London. Lilly Library, Indiana University. Retrieved via Wikimedia Commons. Public domain. https://commons.wikimedia.org/wiki/File%3AGreat_Palm_House%2C_Kew_Gardens_%28color_engraving%29.png

Figure 11.1: (Left) *Alice's Adventures in Wonderland*, by L. Carroll (J. Tenniel, Illus.), 1865. Public domain, via Wikimedia Commons. https://commons.wikimedia.org/wiki/File:Alice_par_John_Tenniel_25.png (Right) *Through the Looking-Glass, and What Alice Found There*, by L. Carroll (J. Tenniel, Illus.), 1907. Public domain, via Polona Digital Library. https://polona2.pl/item/through-the-looking-glass-and-what-alice-found-there,MTQxMTEoODA4/86

Figure 11.2: *The orphans at their mother's grave* [Spirit photograph stereograph], by Melander & Bro., 1877. From the Library of Congress, Prints and Photographs Division. No known copyright restrictions. https://www.loc.gov/item/2022647932

Figure 11.3: Character.AI. Prompted by Amy Allen and David Hicks. 2025.

Figure 11.4: Character.AI. Prompted by Amy Allen and David Hicks. 2025.

Figure 11.5: ChatGPT. Prompted by Amy Allen and David Hicks. 2025.

Figure 11.6: ChatGPT. Prompted by Amy Allen and David Hicks. 2025.

Figure 11.7: *Into the jaws of death: United States troops wading through water and Nazi gunfire* [Photograph], by R. F. Sargent, June 6, 1944. From the US National Archives and Records Administration. No known copyright restrictions. https://www.loc.gov/item/2021669739

Figure 11.8: ChatGPT. Prompted by Amy Allen and David Hicks. 2025.

Figure 12.1: *Alice, the Zeta Cat and Climate Change*, by M. Boysen, 2016, Potsdam Institute for Climate Impact Research. Used under fair use, via Wikimedia Commons. https://en.wikipedia.org/wiki/File:Alice,_the_Zeta_Cat_and_Climate_Change.jpg

Figure 12.2: *Industrial mill chimneys* [Etching/engraving], by unknown artist, ~19th century. Reproduced via English Fine Cottons. Used under fair use. https://www.englishfinecottons.co.uk/content/uploads/2016/05/Mill-chimneys-etching-1600×949.jpg

Figure 12.3: *Poster calling for the Great Wibsey Low Moor demonstration on behalf of the Ten Hours Bill* [Broadside], by unknown artist, July 1, 1833. Reproduced in The Victorian Web. Public domain. https://victorianweb.org/history/poorlaw/oastler2.html

Figure 12.4: *Photograph of workers in a factory* [Photograph], 1903. From the National Archives (UK) Catalogue ref. COPY 1/501. Open Government Licence v3.0. https://www.nationalarchives.gov.uk/wp-content/uploads/2014/03/copy1-5011.jpg

Figure 12.5: *Frankenstein at Work in His Laboratory* [Illustration], by unknown artist, 1922. From *Frankenstein; or, The Modern Prometheus* by M. Shelley, Cornhill Publishing Company. Public domain. https://commons.wikimedia.org/wiki/File:Frankenstein,_pg_7.jpg

Figure 12.6: Amy Allen and David Hicks. 2025. CC BY-NC-SA 4.0.

Figure B.1: ChatGPT. Prompted by Amy Allen and David Hicks. 2026.

ABOUT THE AUTHORS

Amy Allen

Amy Allen, PhD, is an associate professor of history and social science education in the elementary education program at Virginia Tech. Her research examines how teachers and students navigate history, technology, and inquiry in the age of GenAI. She explores how digital tools can support (not replace) the messy, human work of meaning-making. Her scholarship appears in *Theory & Research in Social Education*, *The Social Studies*, and *Social Studies and the Young Learner*, and she frequently presents at NCSS, NCHE, and AERA. Outside of work, Amy maintains a deeply analog relationship with karaoke, trivia, and chaotic good energy. When she's not researching technocuriosity, she's testing the limits of "place-based learning" at Disney parks, listening to Taylor Swift, and pretending she's in a musical number nobody asked for but everyone secretly enjoys. She is deeply committed to serious scholarship but never taking herself too seriously.

David Hicks

David Hicks, PhD, is a professor of history and social science education at Virginia Tech, where his research explores the integration of digital technologies and media to deepen historical inquiry and disciplinary thinking. With more than ninety publications, he has contributed significantly to understanding how technology can scaffold the analysis of historical sources, foster critical literacy, and humanize digital learning environments. Dr. Hicks's scholarship combines historical rigor, technological innovation, and a commitment to helping teachers navigate the promises and perils of AI in history education. Outside of academia, David remains tragically unrecognized by the Academy, the Recording Industry, and, most cruelly, Mardi Gras king cakes. A self-proclaimed "Gen AI curious" historian, he divides his time between mentoring pre-service teachers, perfecting small animal puppetry, contemplating why Twitter's rebrand destroyed civilization, and surfing TikTok. When not philosophizing about cognitive engineering, David dreams of a One Direction reunion and still insists Blur is better than Oasis.